FLIGHT
MANUAL BY

JEPPESEN

"International Standard Book Number 0-88487-006-5"

J4924B

INTRODUCTION

The *Flight Manual by Jeppesen*, is a complete coverage of the maneuvers required for the private and commercial pilot applicant. It is a study guide for the flight phase of flight training. All the necessary data, from preflight to tiedown, is presented by those who know what it is all about. It is the product of the experience and expertise of numerous flight instructors with thousands of hours of cumulative experience in successfully communicating the basic concepts of "learning to fly."

This manual leads the student through step-by-step discussions with each maneuver liberally explained and well supported by illustrations. Additionally, common misconceptions are dispelled, normally-encountered student errors are discussed in detail, and acceptable performance standards are clearly stated.

FOR THE STUDENT, THIS MANUAL WILL . . .

- insure that he and his instructor are on the same "wave length" regarding the exact techniques and procedures for a given maneuver.

- insure academic understanding of the maneuver prior to the time he is expected to perform in the air.

- act as a source for review at his own leisure when new concepts and procedures are most easily assimilated.

- serve as a reference to good flight techniques long after the coveted pilot certificate is earned.

FOR THE INSTRUCTOR, THIS MANUAL WILL . . .

- act as a ready source for assignments of maneuvers anticipated in the student's next flight lesson.

- prove to be invaluable supplementary material when designing lesson plans for a particular student or situation.

- prove to be an ever-present source for those illustrations so necessary to informative preflight and postflight briefing sessions.

In summary, the *Jeppesen Flight Manual*, if used properly, will be the catalyst which brings student and instructor together in that effort which develops pilots capable of flying with the very best.

TABLE OF CONTENTS

CHAPTER 1

AIRPLANE GROUND OPERATION AND TRAFFIC PATTERNS

SECTION A—PREFLIGHT CHECK AND ENGINE STARTING PROCEDURES

Safe flying *begins on the ground*. The attitudes and habits established in the initial stages of learning to fly will greatly influence the standards that the student pilot will follow throughout his flying career. He should observe his instructor carefully as the preflight operation and engine starting procedures are explained for the first time.

The preflight line check, or visual inspection, is performed prior to each flight. This inspection insures that the aircraft is in a safe condition for the flight. The student pilot should have the normal cautious interest in whether the aircraft is ready for flight or not; plus, as he assumes the duties of pilot in command, *he is responsible* for determining the airworthiness of the aircraft. The flight instructor will point out the various components to be inspected on the particular training aircraft and will explain how to recognize any unairworthy condition.

CHECKLISTS

There are both subtle and major differences between various types and models of aircraft; and the validity of using written checklists for various modes of flight operation has been established by considerable research and experience in both military and civilian aviation.

The use of a written checklist is highly recommended because:

1. It is an organized procedure for a complex operation.

2. The use of a checklist saves time in the "long run."

3. It prevents duplication of effort; or in other words, each item is checked once thoroughly.

4. It insures that no important item will be missed.

5. It helps in making the transition to a different make or model of aircraft.

6. It removes the burden of having to rely on memory.

The preflight inspection is only the first of many procedures that will be carried out according to a written checklist. No matter how many times a procedure is repeated, the checklist should be followed step by step each time.

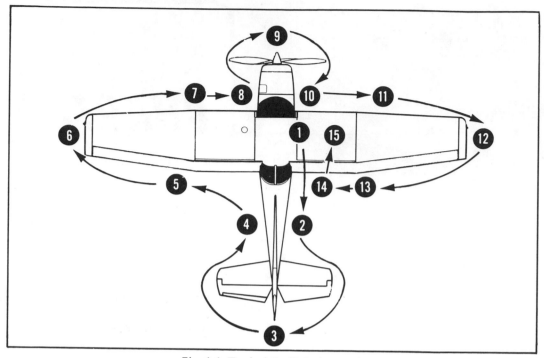

Fig. 1-1. Typical Preflight Inspection

TYPICAL PREFLIGHT

The typical preflight inspection is depicted in figure 1-1 and would proceed as follows:

POSITION ❶ , *CABIN*

The aircraft papers are checked and should consist of

* a. weight and balance data, including equipment list,

★ b. airplane radio station license,

* c. aircraft airworthiness certificate,

* d. aircraft registration certificate,

* e. aircraft flight manual,

    * NOTE: If the aircraft manual is not available, the pilot should

check and determine that the aircraft performance limitations are displayed or placarded.

f. engine and aircraft logbooks (not required to actually be on board the aircraft).

The control lock is removed so that the controls can be checked for freedom of movement.

The fuel quantity guages are checked to determine the amount of fuel in each of the fuel tanks (after turning the master switch ON).

The wing flaps are placed in the full down position. This is done so that the flaps can be more thoroughly inspected during the walk-around inspection.

The fuel valve is moved to the "left tank," "right tank," "both on" position. This is done so that any exces-

* Required by FAA regulation to be on board the aircraft before flight
★ Required by Federal Communications Commission (FCC)

sive water condensation in the fuel lines can drain into the *strainer* during the preflight check.

The **windshield and cabin windows** are checked for cleanliness and general condition before leaving the cabin.

The **instrument panel** is checked for any irregularities, such as cracked glass, and for any instruments or radios which might have been removed for maintenance.

The **magneto switch, master switch, mixture control, and throttle** (MMMT) are placed in the OFF position.

POSITION ❷ , *FUSELAGE*

The **fuselage** is checked for skin wrinkles, dents, and loose rivets. The underside of the fuselage is especially susceptible to rock damage and should also be examined for dents, cleanliness, and evidence of excessive engine oil leakage.

The **static air source** is checked for obstruction, if located on the fuselage. Often, during cleaning or waxing of the aircraft, the small static port becomes plugged with polish or wax. It is important that the static source port be open so that the airspeed, altimeter, and rate-of-climb indicators will function properly.

POSITION ❸ , *TAIL*

The **control surface lock,** if installed, is removed prior to checking the tail assembly. If gusty wind conditions are anticipated, external locks are often used to keep the movable control surfaces from applying heavy loads to the flight control system and to the "stops" that limit the range of control surface movement.

The **fixed tail surfaces** are checked for general condition, such as dents, skin wrinkles, and loose rivets. The underside and leading edge of the horizontal stabilizer (or stabilator) is especially prone to damage by rocks thrown up during takeoffs and landings on unimproved runway surfaces.

The **movable tail surfaces** are checked for damage, loose hinge bolts, and freedom of movement. The control surface stops should be inspected for damage and the surface skin checked for dents and wrinkles.

The **trim tab** is inspected for security and general condition.

The **tail and beacon lights** are checked for damage.

The **navigation antenna** is checked for damage.

Tiedown, chains, or ropes are disengaged.

If the aircraft is equipped with a tailwheel, inspect the steering arms, cables, and springs for wear. Also, check the tire for wear, proper inflation, cuts, and bruises.

POSITION ❹ , *FUSELAGE*

Repeat position 2

POSITION ❺ , *LEFT WING FLAP AND AILERON*

The **control surface locks,** if installed, are removed.

The **wing flap** is completely inspected with particular attention given to all moving parts. The flap hinges are checked for security, wear, and freedom of movement.

The **aileron** is checked for security, damage, and freedom of movement. Also, the aileron pushrods, or control cables, should be checked for security, damage, and tension.

POSITION 6, LEFT WINGTIP

The left **wingtip** is checked for damage and secure attachment.

The left **navigation light** is inspected for damage.

POSITION 7, LEFT WING

The **landing light window**, if applicable, is checked for security, cleanliness, and damage. It should be pointed out that the landing light may be located in positions *other* than on the left wing.

The **wing structure** is inspected. The leading edge is checked for dents and general condition and the surfaces of the wing are checked for the condition of the metal skin and for wrinkles which would indicate internal structural damage. The wings also should be free of mud, frost, snow, or ice that might disturb the flow of air over the wing surface.

The **tiedown** chains or ropes are disengaged.

The **pitot tube** is inspected next. The pitot tube cover is removed, if installed, and the tube opening is checked for obstructions. A plugged pitot tube opening will cause the airspeed indicator to malfunction. Futhermore, the pitot tube itself should show no signs of damage and should not be bent out of alignment with the wing chord line.

The **stall warning vane** on the leading edge of the wing is checked for freedom of movement. It is a good

practice to turn the master switch ON prior to the stall warning vane inspection so that the stall warning signal can be checked when the vane is deflected upward. In the case of a pneumatic stall warning device, the leading edge opening is checked for freedom from obstructions.

The **fuel tank drain** is checked for security and leakage. If the aircraft is equipped with a quick-drain device, drain a few ounces of fuel into a clear container and check for the presence of water. Water can be detected since it is heavier than gasoline and will settle to the bottom of the container and will clearly contrast with the color of the gasoline. If water is found, drain until all the water is eliminated.

The **fuel tank vent** opening is checked for obstruction.

The **fuel quantity level** is checked visually by removing the filler cap and looking into the tank. The quantity in the tank should agree with the fuel gauge reading observed at the beginning of the preflight inspection. At the completion of the fuel level check, the filler cap should be replaced and tightened securely.

POSITION 8, LANDING GEAR

The **main landing gear** is inspected next. The tires are checked for wear, cuts, bruises, and proper inflation. The wheel fairings, if installed, are checked for cracks, dents, and general security. The hydraulic brake and brake lines are visually inspected for leaks.

The **nose gear** is given a thorough visual inspection. Particular attention is paid to the proper inflation of the exposed gear strut, and the nose-wheel tire is checked for wear, cuts, bruises, and proper inflation.

POSITION **9** *, ENGINE AND PROPELLER AREA*

The engine compartment is accessible through the cowl access door. A thorough visual inspection is made inside the engine cowling for loose wires, loose clamps, and for oil or fuel leaks.

The oil quantity then is checked by removing and reading the dipstick. Oil should be added if below the minimum level recommended by the manufacturer. Then, the dipstick is replaced and tightened securely. The oil filler cap is checked for security.

The fuel strainer is drained for several seconds to eliminate any water that may have collected in the fuel strainer. Water will form in the fuel tanks from condensation of moisture in the air; or, in some cases, water may have been present in the gasoline when the tanks were filled.

The front cowl openings are checked for obstructions.

The cowl flaps are checked for security (if the aircraft is so equipped).

The propeller and spinner are checked for security. The propeller blades and tips are checked for nicks and scratches. Propeller nicks of more than approximately one-eighth inch in depth can cause excessive stress in the metal of the propeller and should be filed, or "dressed down," prior to flight by a qualified FAA licensed mechanic.

If the propeller is of the constant-speed type, it is checked for oil leakage. Oil leakage will generally show up as streamers along the propeller blade.

In cold weather, the propeller should be "pulled through" (rotate the propeller) two or three revolutions. This procedure loosens the congealed oil and makes starting easier.

POSITION **10** *, CABIN EXTERIOR*

The windshield and cabin windows are checked for cleanliness and general condition.

NOTE: A dry rag should *not* be used to clean windshields. Rubbing a dry rag *will scratch the windshield* and build up a static electrical charge that causes dust and dirt to cling to the windshield. A cleaner specifically designed for aircraft windshields should be used.

The communication antenna is checked for general condition and security.

POSITION **11** *, RIGHT WING*

The checks of the wing structure, fuel tank drains and vents, fuel quantity, and release of wing tie-downs are performed in the same manner as described in the listing for position **7**

POSITION **12** *, RIGHT WINGTIP*

The procedures outlined for position 6 are repeated here.

POSITION **13** *, RIGHT AILERON AND WING FLAP*

The procedures detailed for position 5 are repeated at this position.

POSITION **14** *, BAGGAGE DOOR*

After loading the baggage, the baggage door is closed and checked for security.

After the preflight is completed, the pilot is ready to begin the prestarting checklist.

STARTING PROCEDURES

Since there are a number of different procedures used to start aircraft engines, the use of an appropriate written checklist to start the airplane is suggested. Although the starting procedure may be different from airplane to airplane, there are certain common safety precautions and suggestions that should be observed.

A pilot should avoid starting the engine with the tail of the airplane pointed toward parked automobiles, spectators, or toward an open hangar door. In addition to being discourteous, it subjects persons to injury, wind blast and debris, and property to possible serious damage.

Prior to starting an airplane on an unprepared surface, the ground under the propeller should be inspected for rocks, pebbles, dirt, mud, cinders, or any other loose particles that might be picked up by the propeller and hurled rearward. Any such particles should be removed as they will *inflict damage to the propeller* as well as to other parts of the airplane. If this is not possible, movement of the aircraft to another position prior to engine start may be required.

The flight instructor will explain the proper use of the engine starter and any special procedures for the engine that must be taken into consideration in starting. Proficiency in starting the engine will come with practice and experience; however, the safety precautions and courtesy requirements listed above essentially remain the same.

An aircraft engine is easy to start if the pilot follows the few basic steps. Figure 1-2 is an example of an engine starting checklist which is typical of those found in the aircraft owner's handbook. The

pilot should follow the recommendations of the aircraft manufacturer for the best engine starting method.

STARTING THE ENGINE

(1) Carburetor Heat — Cold.

(2) Mixture — Rich.

(3) Primer — As Required.

(4) Throttle — Open ¼ inch.

(5) Master Switch — "ON".

(6) Propeller Area — Clear.

(7) Ignition Switch — "START".

(8) Oil Pressure — Check.

Fig. 1-2. Typical Starting Check List

A typical starting sequence for a training aircraft is given in the following paragraphs. It should be noted that the engine controls are used in the same sequence as outlined in the checklist shown in figure 1-2.

1. First, the carburetor heat control should be in the COLD position (full in). (See Fig. 1-3.) By placing the control in this position, the air entering the engine is filtered and dust and dirt are prevented from entering the engine.

2. Next, move the mixture control to the FULL RICH position (full in) as shown in figure 1-4.

Fig. 1-3. Carburetor Heat in Cold Position

Fig. 1-4. Mixture Control In Full Rich
Position

3. The primer is now used to pump fuel into the engine cylinders to aid in starting. (See Fig. 1-5.) The number of primer strokes required depends on how long the engine has been shut down and how cold the weather is. If the engine has been shut down for less than an hour, it will probably start without priming. The recommended procedure, out-

Fig. 1-5. Engine Primer

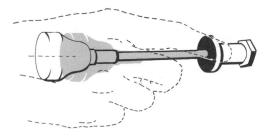

Fig. 1-6. Approximate Throttle Setting
For Starting

lined in most light airplane owner's manuals, is to use from two to six strokes of the engine primer — the greater number of strokes required when temperatures are colder.

4. Now, open the throttle one-quarter of an inch. (See Fig. 1-6.) By opening the throttle only a small amount, additional fuel will be drawn into the engine cylinders. Then, when the engine starts, it will be operating at a low speed which lessens engine wear.

5. Next, place the master switch in the ON position to supply electrical energy to the starter motor. (See Fig. 1-7.)

6. Now, open a window or door and shout "clear" to warn anyone near the aircraft that the propeller is about the rotate. Listen for the usual reply of "clear" from persons in the area, and look around to insure that there is no one in the immediate area.

7. When you are assured that the area is clear, rotate the ignition switch to the START position. (See Fig. 1-8.) When the engine "fires," return the switch to the BOTH position. It may be necessary to "pump" the throttle slightly as the engine fires to keep it running until it is operating smoothly. Excessive

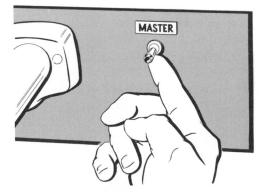

Fig. 1-7. Master Switch ON

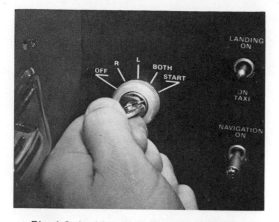

Fig. 1-8. Ignition Switch In Start Position

Fig. 1-9. Oil Pressure Gauge

pumping should be avoided as it is easy to "flood" the engine at this stage.

8. After the engine is running smoothly, adjust the throttle to operate the engine between 800 and 1,000 r.p.m. A low power setting is recommended to prevent undue friction within the engine before the lubricating oil has had a chance to thoroughly coat the engine's internal parts. Simultaneously, check to be sure that the oil pressure gauge is in the green arc range. (See Fig. 1-9.) If the oil pressure does not register properly within 30 seconds in the summer and 60 seconds in the winter, the engine should be shut down to prevent possible damage and to determine the exact nature of the problem. The reason for the longer time interval during the winter is due to the fact that it takes longer for congealed oil to reach the oil pressure sensing bulb.

SECTION B—TAXIING, ENGINE SHUTDOWN, AND TIEDOWN PROCEDURES

TAXIING

The practice and familiarization with taxiing techniques usually begins on the first training flight. One benefit of taxi practice is that a "feel" for throttle usage can be developed. In addition, the pilot can develop familiarity with proper control positioning that will help in crosswind takeoff and landing practice later in the training program. In modern aircraft equipped with tricycle landing gear and a steerable nosewheel, taxiing is relatively easy. However, as with most new experiences, there are some precautions that should be observed.

USE OF THE THROTTLE

The proper use of the throttle during all phases of flight starts during taxi practice when the instructor emphasizes that the throttle is *never* to be used abruptly or with jerky or jamming motions. Erratic throttle movement can cause the engine to falter or not respond quickly. The throttle may be applied rapidly, but it should be made with a smooth motion so the engine will respond without hesitation.

Because some of the knobs and controls in an automobile, as well as the knobs on the instrument panel of an airplane, operate by pulling them *out* to the ON position, some students erroneously assume that the throttle should be pulled out to increase power. However, just as the accelerator pedal on a car is pushed *in* to increase power, so should the throttle in the airplane be pushed *in* to increase engine r.p.m.

A knurled friction knob, such as shown in figure 1-10, is incorporated on the throttle assembly of most light aircraft. This control can be rotated to adjust the

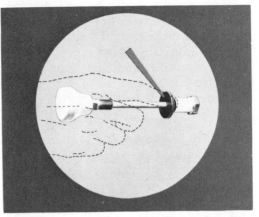

Fig. 1-10. Throttle Friction Knob

friction resistance so that natural arm movement will not overcontrol the throttle nor require excessive pressures to advance or close it. A further assist in making power changes is to place the *index finger* against the friction knob so that smooth, precise adjustments of the throttle can easily be accomplished.

ENGINE COOLING DURING TAXIING

Many of the aircooled engines in use today are closely baffled and tightly cowled. Because of the slow speeds associated with taxiing and ground operations, only small amounts of cooling air are forced through the engine cylinders. Therefore, prolonged ground operations, particularly in warm weather, can cause overheating problems in the engine cylinders even before the oil temperature gauge indicates a pronounced rise in temperature. If installed, the *cylinder head temperature gauge* should be used to monitor engine temperature.

It is advisable to avoid prolonged periods of engine idling and to keep ground operations to a minimum. The aircraft owner's manual recommends proper power settings for engine warmup to provide the optimum flow of cooling air through the engine compartment.

R.P.M. DURING TAXI

More power is required to *start* the aircraft moving than is required to *keep* it moving. Power should be added slowly until the aircraft starts rolling and then reduced in proportion to the desired taxi speed. Naturally, the amount of power required to start and sustain an aircraft in motion is greater on a soft surface than on a hard dry surface.

Taxi speed is *controlled primarily* with throttle and *secondarily* with brakes. Only when a reduction of engine r.p.m. is not sufficient to slow the aircraft should the brakes be used. The use of high power settings and the control of aircraft direction and speed with brakes causes excessive wear and overheating of the braking system. This procedure is as *inappropriate* as the operation of an automobile with a constant foot feed pressure and control of the auto's speed by the application of brakes.

While taxiing, the alternator output light or ammeter should be monitored. If the r.p.m. is too low, the alternator will not develop enough electrical power to supply the electrical needs of the aircraft electrical equipment, especially the radios. If the alternator "low output" light is on, or the ammeter is showing a discharge, it means that the battery is supplying some of the electrical energy being used by the aircraft equipment. Depending on the condition of the battery, there may not be enough power to operate the radios properly.

TAXI SPEED

Many persons who begin flight training are accustomed to the generous stopping capability of the relatively *large four-wheel* brakes on their automobiles. It is helpful to remember that a light aircraft has brakes on only *two wheels* and that the brakes are relatively *small*. Therefore, taxi speeds need to be reduced and held within the capability of the braking

system. Many flight instructors recommend the use of the "rule of thumb" that *taxi speed is equal to a brisk walk.* When operating in confined areas, however, the speed should be slower than the suggested "rule of thumb." A guideline for speeds in confined areas would be that if the brakes failed, the speed should be such that upon reduction of power, the aircraft will, of its own accord, stop short of any immediate obstructions.

Another useful suggestion for the development of proper taxi speed control is for the student to assume that "he has no brakes." If such an assumption is made, the development of proper use of power to control taxi speed will be enhanced and the brakes will be used only as necessary.

TAXI STEERING

Steering is accomplished through the rudder pedals. On most aircraft, the nosewheel is linked to the rudder pedals. When the right rudder pedal is depressed, the nosewheel turns to the right causing the aircraft to turn to the right. Conversely, if the left rudder pedal is pushed, the nosewheel turns to the left causing the aircraft to turn to the left. The angle that the nosewheel can be turned varies between different makes and models of aircraft.

If the pilot desires to make a turn of smaller radius than can be accomplished through nosewheel steering alone, the rudder pedal should be fully depressed in the direction of turn, followed by application of the individual toe brake on that rudder pedal. This procedure will apply differential braking action in the direction of the turn and produce a smaller radius of turn. It is possible to make very tight turns by pivoting on one wheel while using heavy differential braking and a large amount of power. However, this is considered *poor pilot technique* and causes excessive tire wear.

When the rudder pedal is depressed the rudder also moves; therefore, airflow over the rudder from taxi speed, wind, and propeller slipstream will provide a small assist in turning the aircraft. (See fig. 1-11)

EFFECTS OF WIND ON TAXI TECHNIQUE

Taxiing in light or calm winds at moderate speeds does not require any additional skills or use of other controls. However, when wind speeds are moderate or strong, special techniques must be employed. When strong winds are present, there is a tendency for the wind to get under the upwind wing and tip the aircraft toward the downwind side. This tendency can be counteracted by proper placement of flight controls.

It must be understood that at *slow* speeds the aileron, rudder, and elevator controls are relatively *ineffective*. As the speed of airflow over the controls increases, control effectiveness also increases. There are three considerations which cause airflow over the control surfaces: taxi speed, wind speed, and propeller slipstream. Any one, or any combination of the three, may act on a control surface at a given time.

The aircraft controls respond the same whether taxiing at *5 miles per hour in a calm wind* or *standing still with a 5 mile-per-hour headwind*, and the controls will be equally effective in both cases. Also, if the aircraft is taxied at 15 miles per hour *into* a 15 mile-per-hour wind, the controls have a *30-mile-per-hour airflow* over them and respond to that velocity of airflow.

On the other hand, if the aircraft is taxied over the ground at 5 miles per hour with a *tailwind* of 5 miles per hour, the taxi speed and the wind speed are cancelled and the controls, in effect,

Fig. 1-11. Rudder Helps Turn

respond as though *no wind* existed. However, if the aircraft is slowed to a stop, the controls will respond as though there was an increasing tailwind component. When the airplane is stopped completely, the control surfaces will be subjected to the direct effects of a 5 mile-per-hour tailwind.

A study of the necessary control positions required when taxiing in strong winds will enable the pilot to effectively *compensate* for the adverse affects of these winds.

TAXIING IN HEADWINDS

When an aircraft is taxied directly into a headwind, the wind flows over and under both wings equally and has no tendency to tip the aircraft. (See Fig. 1-12.) Under these conditions, the elevator control should be held near neutral or slightly forward of neutral to cause the nose gear to have the normal amount of weight exerted on it.

If the control wheel is held full forward in a strong headwind, downward deflection of air by the elevators forces the nose down and places *more* than the normal weight on the nose gear which compresses the nose strut and brings the propeller tips closer to the ground, as shown in figure 1-13. Normally this is not hazardous, but on rough terrain this procedure could cause the propeller to contact the ground and incur damage. Therefore, the recommended technique when taxiing into a strong wind is to place the *elevator control near neutral* or slightly forward of neutral.

When *taxiing over rough ground* into a strong headwind, it is recommended that the control wheel be held back, so that the elevators are up. This procedure will force the tail down and increase the prop tip clearance. In a strong wind, or with a *strong* prop blast, the nosewheel may come completely off the ground. This is usually not a hazard because the tail skid will protect the tailcone from damage. However, this procedure should be reserved *only* for taxiing over rough terrain. (See Fig. 1-14.)

TAXIING IN TAILWINDS

When the pilot is taxiing the aircraft in a strong tailwind, the control wheel should be placed in the full forward position. This causes the wind to strike the upper surface of the elevator and stabilizer (or stabilator) and exerts a downward force

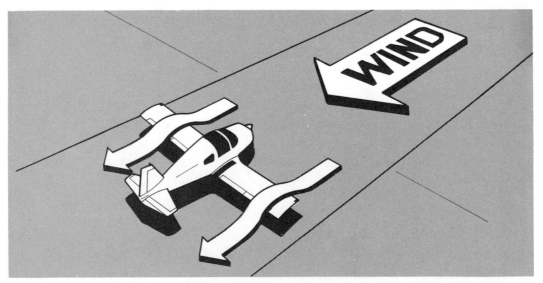

Fig. 1-12. Taxiing In A Headwind

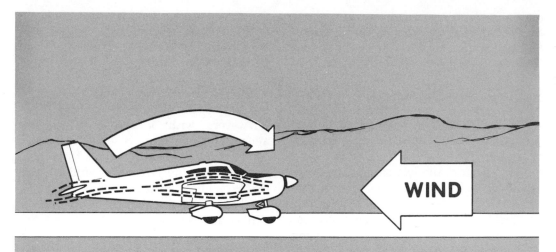

Fig. 1-13. Forward Control Wheel Increases Weight on Nose Gear

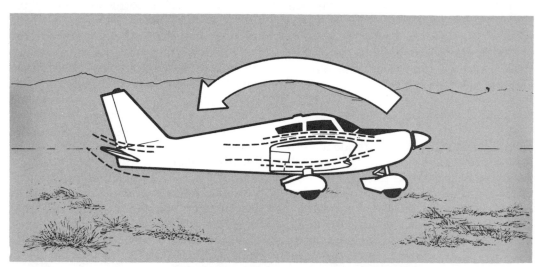

Fig. 1-14. Control Wheel Lightens Weight on Nose Gear

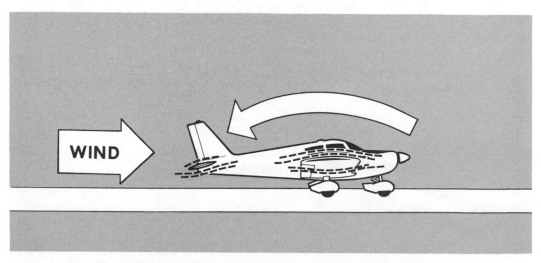

Fig. 1-15. Down Elevator Counteracts Tailwind Overturn Tendency

on the tail. This procedure will prevent the wind from "getting under the tail" and causing the aircraft to noseover. (See Fig. 1-15.)

TAXIING IN CROSSWINDS

When the wind is striking from the left or right of the nose, a lifting effect on one wing will be noticed. With the wind blowing against the left side of the nose, as shown in figure 1-16, the wind pushes against the fuselage and in effect rolls the airplane to the right. This exposes the underside of the left wing to a lifting effect from the wind. The fuselage partly blocks the airflow to the right wing and a pronounced rolling tendency is produced. Normally, the wind is not strong enough to cause the aircraft to actually tip, but with a strong wind and lacking correct control placement, the aircraft could be upset.

QUARTERING HEADWIND

To counteract the tipping tendencies of a left quartering headwind, the controls should be positioned as follows: the control wheel should be turned full left. The wind flowing over the left wing will exert pressure against the up aileron, as illustrated in figure 1-17, tending to force that wing down. The airflow over

the right wing will apply pressure to the down aileron, tending to hold the right wing up. This combination of forces acts to balance the tipping or rolling effects.

It should also be noted that the stronger the wind, the greater the tipping force. However, with a strong wind, the ailerons are also more effective which tends to compensate for the tipping effect of the wind. With the wind quartering from ahead of the airplane, the elevators are used in the same manner as when going directly into a headwind.

Throughout this discussion, when the phrase "moving the aileron controls" is used, the intent is to move the controls all the way to the *stops* so the ailerons are fully deflected. At slow taxi speeds, the controls are less effective so a small amount of deflection may be of little value in affecting proper control.

When the wind is quartering from ahead and from the right, the tipping tendency is to the left, as shown in figure 1-18. In this situation, the control wheel is turned full right so the right aileron is up and the left aileron is down. The ailerons in this position again provide a restoring force to counteract the tipping force while the elevators are used as if taxiing straight into the wind.

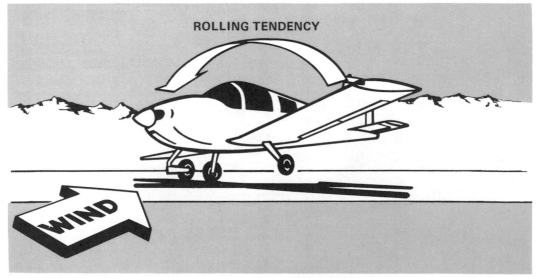

Fig. 1-16. Quartering Headwind Effect.

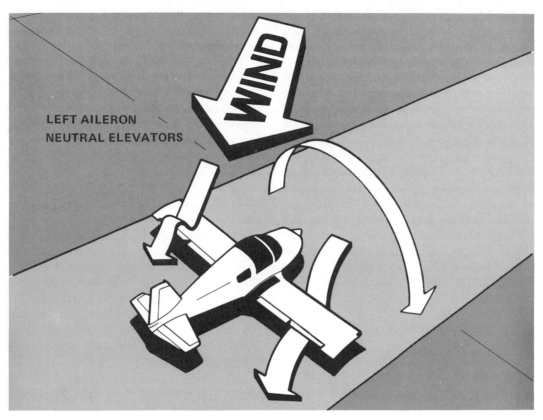

Fig. 1-17. Control Positions For Left Quartering Headwind

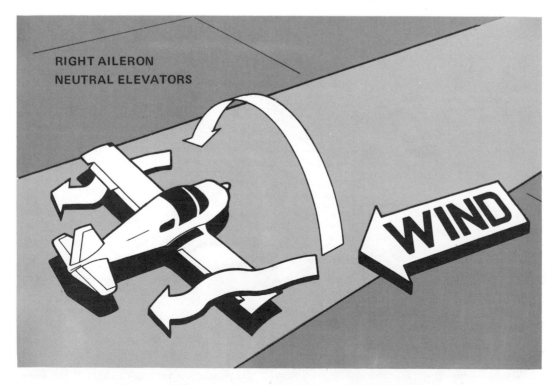

Fig. 1-18. Control Position For Right Quartering Headwind.

In order to condense this discussion into an easily recalled rule, remember that when the crosswind is quartering *from ahead* of the aircraft, the pilot should *steer into the wind* or turn the control wheel "into the wind." Putting the aileron control "into the wind" is an expression heard frequently throughout flight training.

Taxiing with a *quartering headwind* is exactly the same condition encountered in a crosswind takeoff or landing approach and rollout. Understanding the effects of wind and the development of the technique and feel for the airplane in a quartering headwind can be directly transferred to later takeoff and landing practice.

QUARTERING TAILWIND

When the wind is striking an airplane from behind, the ailerons must again be positioned to counteract the tipping tendency. In a quartering tailwind situa-

tion, the control wheel must be rotated *away* from the wind. When the wind is from the *left rear*, referred to as a left quartering tailwind, the aileron control must be turned to the *right*, as shown in figure 1-19. This results in a downward aerodynamic force being applied to the downturned left aileron and an upward force on the right aileron, thereby counteracting the tipping tendency. (See Fig. 1-19.)

With a right quartering tailwind, the aileron control is turned to the left. Therefore, the rule to remember when taxiing with a crosswind from behind the airplane is to turn the ailerons "out of the wind" or away from the wind. Quartering tailwinds also have a tendency to get under the elevator (or stabilator) and lift the tail. When this happens, the reaction is to tip the aircraft up on the nosewheel and one main wheel. To counteract this tipping force, the elevator control is pushed forward so the elevators are in the DOWN position. The tail-

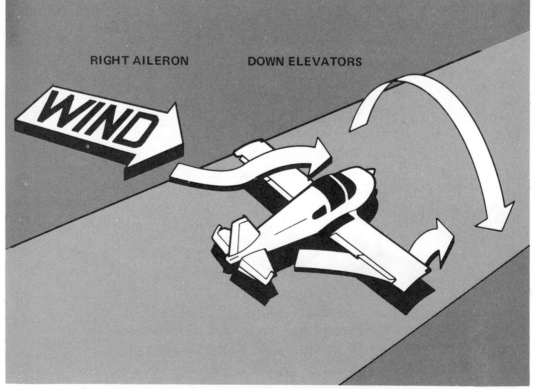

RIGHT AILERON   DOWN ELEVATORS

WIND

Fig. 1-19. Control Position For Left Quartering Tailwind

wind will then apply an aerodynamic force on the top of the elevators tending to push the tail down, as previously shown in figure 1-15.

When taxiing in a strong wind, there is one situation in which the pilot must be particularly cautious; that is, *slowing down with a quartering tailwind while turning into the wind.* As shown in figure 1-20, the increasing tailwind component combined with the normal tendency of the airplane to tip outward makes the aircraft especially vulnerable to being "upset." Slow taxi speeds and slow turns minimize the overturn danger.

During much of the time that the student spends taxiing, the winds will probably be at a level where control positioning will not be especially critical. However, it should be evident that determining the wind direction with respect to the aircraft heading and proper control placement requires practice.

PRACTICING PROPER CONTROL POSITIONING

It is recommended that proper control positioning be practiced on *every* flight. When there are light winds, it should be *assumed* that the winds are 30 miles per hour and from whichever direction the windsock is pointing. Then, correct placement of the aileron and elevator controls can be conscientiously practiced as if they were definitely needed. The day when there are strong winds which require proper control positioning is not the time to be hesitant or unsure about procedures.

The benefits of this practice will be evident later in the flight training program. First of all, the student pilot will be prepared for the day when he actually encounters strong winds. Equally important, he will develop a constant awareness of wind direction early in the program. Also, by the time the aircraft has reached the takeoff position on any given flight the student pilot will have the wind speed and direction well evaluated. He will be able to determine what crosswind takeoff and landing techniques are required before reaching the runup area. Furthermore, the proper control usage for crosswind taxiing will enable the student to be well prepared for extensive crosswind takeoff and landing practice.

HAND SIGNALS

Instruction in taxiing should include familiarization with the standard hand signals used by ramp attendants for directing pilots during ground operations. Figure 1-21 illustrates these signals. The

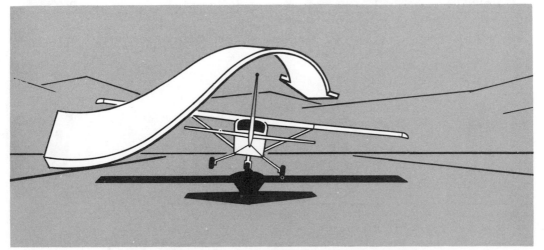

Fig. 1-20. Turning With A Tailwind Requires Caution

Fig. 1-21. Hand Signals

beginning pilot should study these signals carefully prior to his first instructional flight and review them periodically during the first few hours of training.

ENGINE SHUTDOWN AND PARKING PROCEDURES

Upon the completion of a flight, the aircraft should be taxied into the parking area utilizing a safe taxi speed and the control procedures discussed in the taxi section of this chapter.

Upon reaching the desired parking area, the pilot should proceed with the engine shutdown according to the checklist found in the owner's manual. A typical checklist is shown in figure 1-22, and is explained as follows:

1. First, stabilize engine r.p.m. at approximately 1,000 r.p.m.

2. Next, turn the radios OFF.

3. If the aircraft is equipped with auxiliary or boost pumps, they should be turned OFF.

4. Now, pull the mixture control to the IDLE CUTOFF position.

5. Next, turn all accessory switches OFF.

6. Then, place the ignition and master switches in the OFF position.

7. Finally, install the control lock.

ENGINE SHUTDOWN

(1) RPM -- 1000

(2) Radio (s) -- OFF

(3) Boost Pumps -- OFF

(4) Mixture -- IDLE CUT-OFF
(pulled full out)

(5) Accessory Switches -- OFF

(6) Ignition and Master Switches -- OFF

(7) Control Lock -- INSERTED

Fig. 1-22. Engine Shutdown Check List

MOVING AIRCRAFT BY HAND

Sometimes it becomes necessary to move the airplane by hand. This is usually the case when "hangaring" an airplane.

Since the structure of an airplane is composed of lightweight materials, caution must be observed in *where* to apply pressure for pushing or pulling. There are certain "do's" and "don'ts" that should be observed. For example, only light, general aviation aircraft should be moved by hand. If the aircraft is a single-engine, four-place or less, it can probably be moved by one person without too much difficulty. Larger aircraft should be moved by a tow tractor or other powered tow equipment. If powered vehicles are unavailable, then several people should participate in pushing the "big bird."

Most tricycle-geared light aircraft are easily and safely maneuvered by a towbar attached to the nosewheel. (See Fig. 1-23.) The towbar is usually stowed in the aircraft baggage compartment and should be used when available. The airport fixed base operator usually has an assortment of towbars for use if none is found with the aircraft.

When using a towbar, it is necessary to ensure that the nose gear is not turned *beyond* its swivel limits. There are limit stops on some aircraft which restrict the gear from turning beyond a given arc. Others have limit marks painted on the gear to indicate the travel range permitted. Turning the nosegear in either direction beyond its steering radius limits will result in damage to the nose gear and steering mechanisms. The owner's manual should be consulted for this information.

It is difficult to judge the proximity of wingtips and tail sections when pushing the aircraft aft and, at the same time, control the direction by hand with a towbar attached to the nosewheel.

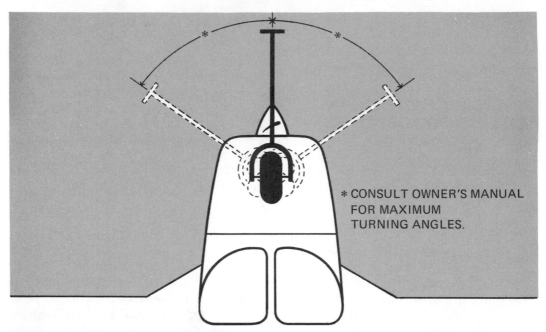

*CONSULT OWNER'S MANUAL FOR MAXIMUM TURNING ANGLES.

Fig. 1-23. Nose Wheel Turn Limits

Bumping the tail against the rear wall of the hangar will, at best, cause minor damage (often referred to as "hangar rash"). Therefore, when moving airplanes alone and by hand, caution is the "watchword." Many owners place "two-by-four's" or similar size wood blocks on the hangar floor to stop the main wheels as the plane is pushed into the hangar.

If a towbar is not available, a few tips for maneuvering the aircraft are listed as follows:

1. **Pivoting the aircraft** — a tricycle-geared aircraft can be pivoted about one main wheel if the nose gear is lifted off of the ground. Applying downward pressure at the tail usually will "do the job." Care must be exercised where pressure is applied, however. (See Fig. 1-24.)

   For example, pressure should be applied over a bulkhead just ahead of the vertical fin. (See item 1.) The tail of an aircraft having a horizontal stabilizer may be depressed by applying hand pressure on the front spar near the fuse-

lage. (See item 2.) Care should be observed to never lift or push on the *outer ends* of the horizontal stabilizer or elevator. Stabilators should not be used as push points on aircraft equipped with this type tail surface. With the tail depressed, the nosewheel will clear the ground and the aircraft can readily be turned in either direction by pivoting it about one of the main wheels. (See item 3.)

The aircraft wingtips should be carefully observed because one wingtip moves aft while the other moves forward. Structural damage may result if a wingtip strikes another object.

Pivoting a tailwheel-type airplane is quite easy because the tailwheel usually will pivot 360°. Pushing on the side of the fuselage (over a bulkhead) adjacent to the horizontal stabilizer permits one person to turn the airplane around. (See Fig. 1-25.)

2. **Pushing the airplane** — Pushing on the leading edge of the horizontal

1. APPLY PRESSURE OVER BULKHEADS

2. PUSH DOWN ON FRONT ELEVATOR SPAR NEAR FUSELAGE

3.

Fig. 1-24. Pivoting Tricycle-Gear-Type Airplane By Hand

stabilizer is acceptable if caution is exercised. Pressure should be applied only at rib locations and near the fuselage, as shown in figure 1-26. Up or down loads should be avoided since the leading edges of light aircraft wings and tail surfaces are usually manufactured of light gauge material and damage could result.

The leading edge of the wing is also a good push point on low-wing aircraft. Again, pressure must be applied only at rib locations. The disadvantage of using the wing is that more than one person is needed to maintain directional

Fig. 1-25. Pivoting Tailwheel-Type Airplane

Fig. 1-26. Push At Stabilizer Rib Locations

control of the aircraft. On high-wing aircraft, wing struts make good push points. (See Fig. 1-27.) Again, two persons are required to easily move the airplane. The propeller should *never* be pushed or pulled *near the tip*. If pressure is applied to the propeller tip, it may be permanently bent out of "track" and produce a serious vibration problem in flight. If pushing on the propeller is necessary, it should be done *near the hub*. The

nosecap of the engine cowling and the trailing edges of wings and control surfaces are not designed for the type of load incurred and *should not* be used to move the airplane.

AIRPLANE TIEDOWN

Proper tiedown procedures are the best precaution against damage to a parked airplane by strong or gusty winds. Once the aircraft is in the proper tiedown area,

Fig. 1-27. Wing Struts Provide Good Push Points

TIEDOWN EQUIPMENT

WING TIEDOWN

AILERON LOCKS

RUDDER LOCK

N7093M

CONTROL LOCK

TAILCONE TIEDOWN

Fig. 1-28. High-Wing Airplane Tiedown Procedures

the parking brakes should be released (if set) and chocks placed in front and behind the main wheels, as shown in figure 1-28.

In addition, all flight controls should be locked or tied to prevent the control surfaces from banging against the stops. Some aircraft are equipped with internal control locks operable from the cockpit. On others it may be necessary to use external padded battens (control surface locks) or secure the control wheel and rudder pedals inside the cockpit. (See Figs. 1-28 and 1-29.) Even if the airplane is equipped with internal control locks, the additional use of external locks is a good practice because they prevent gust loads from being fed back into the control system. The ailerons and rudder should be secured in the neutral position.

When using external control surface locks, it is advisable that red streamers, weights, or a line to the tiedown anchor be fastened to the lock. This provides a means of alerting the pilot and airport ramp personnel to remember to remove the external locks prior to flight.

Tailwheel-type aircraft headed into the wind should have their elevators secured in the UP position by securing the control column in the full aft position. On the other hand, tailwheel-type aircraft "tailed" into the wind should have their elevators in the DOWN position by securing the control column or stick in the full forward position. Tricycle-gear type aircraft should have the elevators secured in the neutral position.

The wing tiedown should be accomplished with nylon or dacron tiedown ropes or chains capable of resisting a pull of approximately 3,000 pounds. They should be secured without slack, but not tight. Tiedown ropes actually put *inverted flight stresses* on the aircraft, and many aircraft are not designed to take excessive inverted loads. For this reason, too much slack will allow the aircraft to jerk against the ropes. On the other hand, too little slack may permit high stresses to be placed on the aircraft.

Tricycle-geared aircraft should be secured at both the nose gear and tailcone. When securing the nose gear, care should be exercised to hook to tiedown provisions on the nose gear and protect it from abrasive damage from chains or tiedown ropes. The attachment to the nose gear and tailcone should be made so that side movement of the aircraft is restricted. Most damage to tied down aircraft takes place because of loose tiedown chains or ropes which permit the aircraft to jerk against the restraints.

Figures 1-28 and 1-29 show proper vertical chain tiedown procedures. One link on the free end of the chain is passed through a link of the taut portion and a safety snap is used to keep the link from passing back through. Therefore, any load on the chain is borne by the chain itself instead of the snap.

The tail tiedown should be taut *but not so tight as to raise the nose of the airplane* since, in a headwind situation, this will increase the angle of attack of the wing and create additional lifting force, thereby causing more pressure on the wing tiedown restraints.

A mass tiedown arrangement used by many airport operators consists of continuous links of parallel wire ropes pas-

Fig. 1-29. Low-Wing Airplane Tiedown Procedures

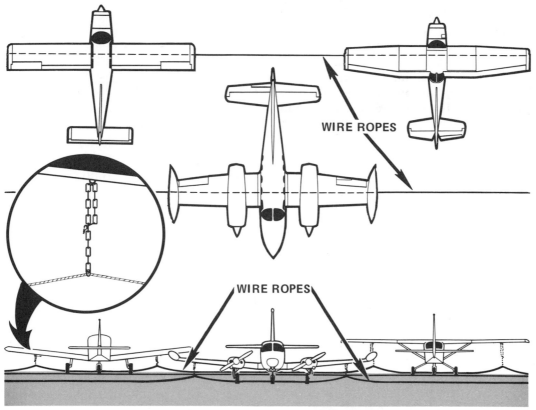

Fig. 1-30. Typical Wire Rope Aircraft Tiedown Systems

sed through U-bolt anchors, secured in cement within the ramp surface. Tiedown chains are attached to the wire rope with roundpin galvanized anchor shackles. This allows the tiedown chains to "float" along the wire rope and provide a variable distance between anchor points so that a variety of large, medium, and small aircraft can use the vertical tiedowns without loss of space. A top view and front view of a typical wire rope tiedown system is shown in figure 1-30. The correct tiedown procedure used at your airport will be explained by your instructor.

CLEANUP

Once the aircraft has been secured, the pilot should make a careful check to be sure all switches are in the OFF position; any trash, papers, or flight planning items should be cleaned from the cockpit area; the pitot tube cover should be installed if applicable; and the propeller placed in a horizontal position to lessen the possibility of damage caused by another taxiing aircraft's wingtip. Placement of the propeller in a horizontal position also permits the spinner to offer more protection to the prop hub from rain, snow, and ice. This is especially important for an aircraft equipped with a constant-speed propeller.

SECTION C—TRAFFIC PATTERNS

Since the new student will fly in the airport traffic pattern on the first training flight, a general discussion of procedures will enable him to feel more "at home" during the initial flight training sessions. For the beginning pilot, work in the traffic pattern and practice in take-offs and landings is usually enjoyable. Here, he has an opportunity to put a series of individual basic maneuvers together to perform a useful task — the approach and landing.

The FAA regulations regarding traffic patterns are very broad and general. Essentially, they state that an airplane approaching an airport to land will *make all turns to the left*, unless a right traffic pattern is indicated.

In training, it is the practice to define the traffic pattern more explicitly, giving identifying names to the various legs and assigning certain altitudes. This has a two-fold advantage. *First*, it gives the student specific reference points throughout the traffic pattern which he can use to assess his performance. This helps him to improve his judgment and proficiency. *Second*, when a pilot flies the traffic pattern properly, pilots approaching or in the traffic pattern can better predict what other aircraft will do and act accordingly. This contributes to the smooth flow of air traffic.

NORMAL TRAFFIC PATTERN

Figure 1-31 illustrates a normal light aircraft traffic pattern. It is rectangular in shape, has five named legs, and one designated altitude. The discussion on traffic patterns that follows is based upon the type of aircraft generally used in flight training programs and on airports that are *not* served by a control tower. The student will find variations from the described pattern at different localities and at airports *with* control towers. This pattern can be visualized as being placed or overlayed on the runway in active use. Since all turns are normally to the left, this pattern is called a left-hand pattern.

As shown in figure 1-32, a right-hand pattern may be designated to expedite the flow of traffic when obstacles or concentrations of population make the use of a left-hand pattern undesirable.

Since the left-hand pattern is the standard traffic pattern in use, this type of pattern will be used to identify and explain the various traffic pattern legs. An

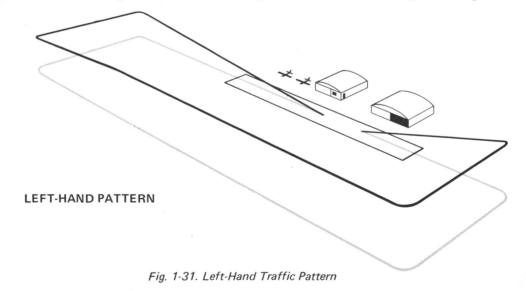

LEFT-HAND PATTERN

Fig. 1-31. Left-Hand Traffic Pattern

RIGHT-HAND PATTERN

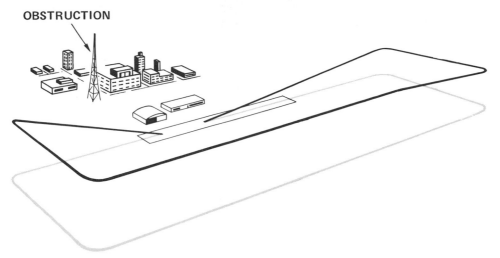

OBSTRUCTION

Fig. 1-32. Right-Hand Traffic Pattern

overhead view of a typical left-hand pattern is shown in figure 1-33 and is explained as follows:

1. **Takeoff Leg or Upwind Leg** — Since the takeoff is normally made into the wind or nearly so, this leg is also oriented into the wind.

2. **Crosswind Leg** — As the name implies, this leg is at a 90° angle to the takeoff leg.

3. **Downwind Leg** — This leg is also appropriately named. The aircraft flies downwind in a direction which is opposite to the takeoff and final approach leg. The downwind leg is *usually 800 feet above ground level*, but it can be designated as 1,000 feet, 600 feet, or some other height, depending on local conditions. The height of the downwind leg above the ground gives the pattern its name. For example, "800-foot pattern."

4. **Base Leg** — This leg is oriented 90° to the downwind leg.

5. **Final Approach Leg** — This leg is aligned with the runway and therefore into the wind.

TAKEOFF LEG

The takeoff, or upwind leg, normally consists of the aircraft's flight path after takeoff, up to the initial climbout altitude of approximately 400 feet AGL. The aircraft should be flown directly above an imaginary extension of the runway centerline and not permitted to drift to one side or the other. No turns should be initiated below 400 feet above the surface or before reaching the airport perimeter.

CROSSWIND LEG

The crosswind leg begins after crossing the airport boundary at an altitude of at least 400 feet above the surface.

DOWNWIND LEG

The downwind leg is flown parallel to the runway and at a distance that would permit a safe landing on the runway should the aircraft experience mechanical difficulties. It should be flown at the designated traffic pattern altitude.

BASE LEG

The base leg begins at a point decided upon after considering other traffic and wind conditions. If the wind is very strong, the turn begins *sooner* than normal. If the wind is light, the turn to base is *delayed*. Other small errors in the downwind leg due to misjudgment in speed, height, and distance from the runway can be compensated for by adjusting or "playing" the turn to base leg.

FINAL APPROACH

The final approach is the path that the aircraft will fly immediately prior to touchdown. It is flown along an imaginary extension of the centerline of the runway and, therefore, the pilot must compensate for any crosswind conditions.

TRAFFIC PATTERN DEPARTURE

When a control tower is in operation, the pilot can request and receive approach for a straight-out, downwind, or right-hand departure. This departure request should be made while asking for takeoff clearance. At airports without an operating control tower, the pilot must comply with the departure procedures established for that airport by the FAA. These procedures usually are posted by airport operators so pilots can become familiar with the local rules.

TRAFFIC PATTERN ENTRY

Traffic pattern entry at airports with an operating control tower is specified by the tower operator. At uncontrolled airports, traffic pattern altitudes and entry procedures vary according to established local procedures. Pilots should familiarize themselves with these procedures and follow local flight instructor recommendations. At unfamiliar airports, airport advisory service or UNICOM, when available, should be utilized for receipt of traffic pattern and landing information.

TRAFFIC PATTERN COURTESIES

On occasion, a number of other aircraft may be in the traffic pattern practicing landings and takeoffs. If they all assume normal spacing, they may completely block the pattern. Other aircraft attempting to enter the pattern or take off will have difficulty acquiring a safe spacing. If a pilot in the pattern sees someone attempting to enter the pattern, he should *extend* his upwind or takeoff leg and allow the other aircraft room for a proper entry.

If a pilot observes someone waiting for an extended period of time for takeoff, he should pass up that landing and go around. The courtesy extended to fellow pilots is greatly appreciated. A thorough discussion of traffic pattern procedures will be covered in the discussion on takeoffs and landings.

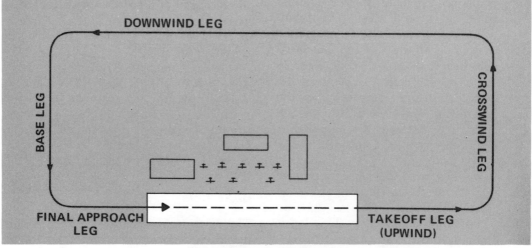

Fig. 1-33. Traffic Pattern Legs

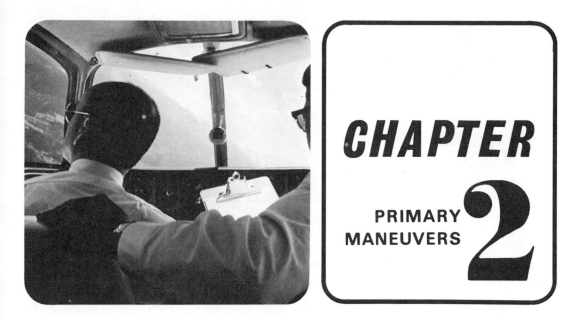

CHAPTER 2

PRIMARY MANEUVERS

SECTION A — STRAIGHT-AND-LEVEL FLIGHT

LEVEL FLIGHT

Flight training normally begins with instruction in the techniques of straight-and-level flight. Experienced pilots acknowledge that no one can maintain perfectly straight-and-level flight; therefore, straight-and-level flight may be defined as a series of recoveries from slight climbs, descents, and turns. The objectives of straight-and-level flight are to point the aircraft in a particular direction, maintain that direction, and fly at a predetermined altitude.

The pilot controls the aircraft's direction and altitude by controlling the nose and wing positions in reference to the natural horizon. This is called *attitude flying*. During attitude flying instruction, it will be learned that there is one fixed-nose position and one fixed-wing position with respect to the horizon for each flight condition. With a constant power setting and the aircraft's attitude adjusted to these fixed positions, the aircraft will maintain the selected flight condition.

There are two other terms with which the beginning pilot should become familiar: *visual flying* and *instrument flying*. Visual flying simply means that the natural horizon outside the aircraft is used as a reference point. Instrument flying is performed when the pilot refers to the flight instruments for wing position and heading reference. This is done with the attitude indicator (artificial horizon) and the heading indicator. (See Fig. 2-1.)

CONTROLLING THE WINGS-LEVEL ATTITUDE

In straight-and-level flight, the wings remain level with the horizon and the fuselage is parallel to the earth's surface. To maintain straight-and-level flight, it is necessary to fix the relationship of the aircraft with the horizon. To do this, the pilot picks a point on the wingtip for a reference point and the wings-level position is maintained by keeping the wing-tips a certain distance from the horizon and parallel to the horizon. Figure 2-2 shows how the wings-level attitude looks

NOSE POSITION   ARTIFICAL HORIZON

STRAIGHT AND LEVEL

FIXED   FIXED

WING POSITION   MINIATURE AIRCRAFT

Fig. 2-1. Visual And Instrument References

in relation to the horizon in a typical low wing and a typical high wing airplane.

CONTROLLING THE PITCH ATTITUDE

To control the pitch attitude, or nose position, the pilot picks some point on the nose or a spot on the windshield as his reference point. This point should be *directly in front of the pilot* rather than over the center of the aircraft's nose. Some instructors assist the student in initially establishing the reference spot by making a mark with a water soluble pen on the inside of the windshield. (See Fig. 2-3.) If the student sights directly over the center of the nose to a point on the horizon, the aircraft will be placed in a skid. Just exactly how the wingtips and nose appear in reference to the horizon will depend upon the type of airplane being flown, how tall the pilot is, and how he positions his seat.

The apparent distance between the horizon line and the nose of the aircraft will change as the pilot changes eye level or seating position. Therefore, it is important that he find a comfortable seating position, sit in an erect manner, and use that same position each time he flies so that the reference point does *not* change.

In level flight, the nose position will appear similar to that shown in figure 2-4. The instructor will put the aircraft in level flight and ask the student to observe the reference point. He will ask the student to note precisely how this point looks from his seat. At this time, the instructor may actually make a mark or "spot" on the inside of the windshield as discussed.

The key element in attitude flying is to discover the attitude for level flight; that is, the wing and nose positions with respect to the horizon. *These positions will always remain the same* as long as the

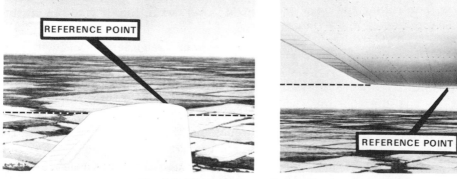

REFERENCE POINT

REFERENCE POINT

Fig. 2-2. Wing Reference Points For Level Flight

Fig. 2-3. Nose Reference Point For Level Flight

Fig. 2-4. Nose Position For Level Flight

pilot sits in the same position in the same type aircraft. They should become so familiar to the pilot that when sitting at home he can close his eyes and project a clear mental picture of the wing and nose positions.

CONTROLLING THE HEADING

In practicing straight-and-level flight, the instructor will ask the student to hold or maintain a certain compass heading and a certain altitude. To do this, the student will set the basic wing and nose attitudes for level flight, then periodically refer to the heading indicator and altimeter to verify that he is on the desired heading and at the preselected altitude.

ATTITUDE FLYING

When flying by instrument reference, the student soon learns the necessity for *scanning*. He should develop the habit of keeping his *eyes moving* continuously between reference points and observing other aircraft traffic in his vicinity. As shown in figure 2-5, the eye should move from one wing, to the nose, to the appropriate instruments, and then to the other wing. At no time should the pilot concentrate entirely on *any one reference*.

Several forces may cause the aircraft to drift from the desired attitude. A power change, slightly below or slightly in excess of the required power setting, will cause the aircraft to descend or climb. Turbulence, wind gusts, and brief periods of inattention to wing position can cause the aircraft to change heading or altitude.

The instructor will direct the student to hold or maintain his assigned heading and altitude. This may give the student the impression that it is possible to rigidly hold an attitude with no allowance for deviation, but this impression is simply not true. Even the most experienced pilots cannot hold the aircraft's attitude constant with no deviation. Flying is a *continuous series of small corrections*.

The necessary corrections should always be made in two steps. *First, stop whatever is happening.* If the heading is changing or the altitude is changing, apply control pressures to return to the level flight attitude. *Second*, adjust the attitude reference points to *make a slow correction back to the desired readings* and change the power setting if required.

Fig. 2-5. Reference Points And Instruments

After the corrections are made, the pilot should return to the normal attitude references. It is important to make slow and easy corrections in order to prevent large overcorrections and "needle chasing" caused by the new pilot's eagerness to perform well.

Some examples of how the aircraft's nose and wing positions look when the attitude is slightly divergent from straight and level are shown in figures 2-6 and 2-7. In figure 2-6, the left wing is down and the nose is level. In figure 2-7, the left wing is high and the nose is high.

Fig. 2-6. Left Wing Down And Nose Level

Fig. 2-7. Left Wing High And Nose High

The nose position helps the pilot tell when the right wing is down. The wing-tip also gives a small clue that the nose is up.

In each position, the pilot's action is to first return to a straight-and-level attitude, then check the appropriate instruments (heading indicator and altimeter) and note the amount of correction necessary. Finally, he should make a very small attitude change to bring the instruments back to the desired readings.

ALTITUDE AND ELEVATORS

The beginning pilot will discover that changes in pitch attitude (elevator) result in altitude changes during cruising flight. Elevators also effect the rate of climb or descent during altitude changes. Since the elevators control the pitch position of the nose, it can be said that altitude is *controlled primarily by the elevators.*

AIRSPEED AND POWER

It will be found that once the student gains reasonable control over the aircraft attitude (hence control of altitude), he will also be able to control airspeed *primarily with power.* (See Fig. 2-8.)

A simple rule of thumb is stated as follows: *when power is variable and available, the throttle is used primarily to control airspeed and the elevators (or stabilator) to control altitude.*

However, after the airspeed has been established at cruise airspeed and power is constant, it is possible to use airspeed variations to reflect changes in pitch attitude. As illustrated in figure 2-9, if the nose is *high*, the airspeed will indicate *slower* than desired. If the nose is *low*, the airspeed will indicate *higher* than desired. If the nose is *moving*, the airspeed will be *changing;* if the nose is *steady*, the airspeed will be *steady*.

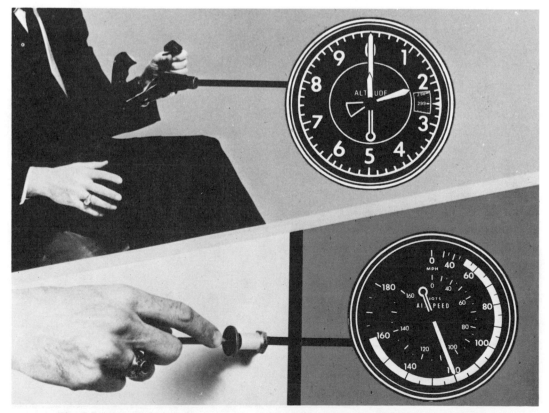

Fig. 2-8. Throttle Primarily Controls Airspeed; Elevator Primarily Controls Altitude

TRIM

To avoid continuously applying forward or back pressure on the elevator to hold attitude, a trim adjustment should be made. Trim is used to make flying easier by removing the need to constantly hold control pressures.

If the airplane feels nose heavy, the pilot is holding back pressure to maintain a given attitude. He should move the trim control to lessen the required pressure until he arrives at a setting which requires *no pressure* to maintain the proper attitude. (See Fig. 2-10.)

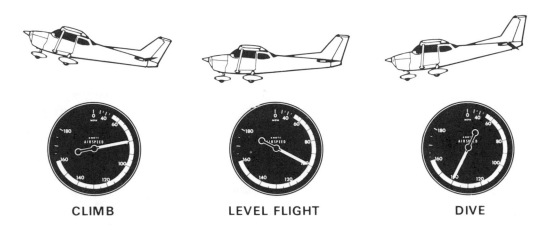

CLIMB   LEVEL FLIGHT   DIVE

Fig. 2-9. Airspeed Is Sensitive To Pitch When Power Is Constant

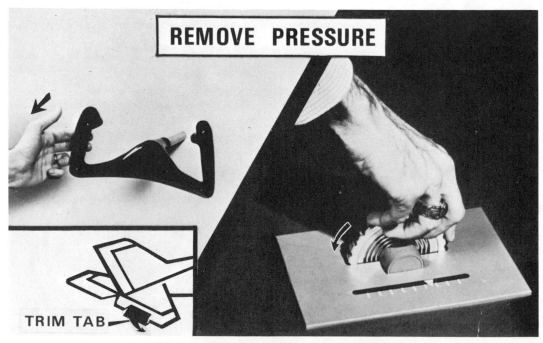

REMOVE PRESSURE

TRIM TAB

Fig. 2-10. Trim Tab Relieves Control Pressures

The trim mechanism is used to *remove* control wheel pressure. It is not used to *fly* the aircraft. The proper procedure is to set the aircraft in the desired attitude and at the selected airspeed and then trim away any control pressure necessary to hold the airplane in that attitude. *Trim tab adjustments should be made* anytime there is a desired *change in airspeed, attitude, or power* in which the aircraft will remain in a new attitude, speed, or power setting even for short periods of time.

SEAT OF THE PANTS FEEL

In time, the new student will develop his *kinesthetic sense*. This is generally defined as the feel of motion and pressure changes through nerve endings in the organs, muscles, and tendons — the feeling pilots describe as the "seat of the pants" sensation. During the student's first flight, he may not have this sense or feel to any great degree, but with practice he will develop the kinesthetic sense.

RUDDER EFFECTS

Aircraft are designed so that in cruising flight they remain in stable aerodynamic balance. Therefore, aileron or rudder pressures are not constantly required. This balanced condition is sometimes referred to as flying "hands off," meaning the pilot can remove his hands and feet from the controls and the aircraft will continue flying straight.

Normally, in flight, the rudder is streamlined with the airflow over the aircraft and the ball of the turn coordinator, or

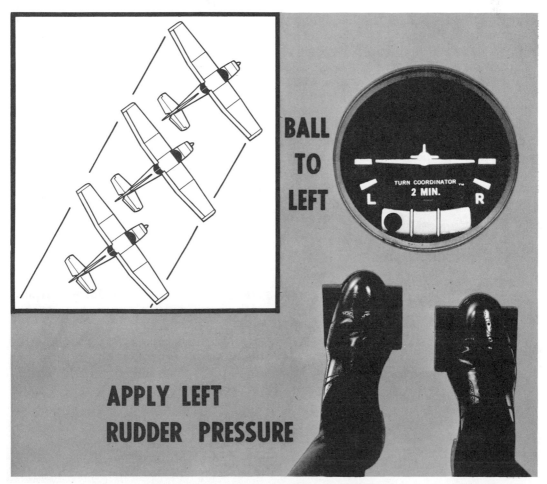

BALL TO LEFT

APPLY LEFT RUDDER PRESSURE

TURN COORDINATOR 2 MIN. L R

Fig. 2-11. Skid Shown With Ball Out Of Center

turn-and-slip indicator, remains in the center of the *inclinometer*. If the ball moves out of the center, as shown in figure 2-11, an unbalanced force on the tail is indicated and the aircraft is skidding. Rudder pressure should be applied to return the ball to the center.

A rule for quick corrections for rudder trim imbalance is: *when the ball is out of the center, "step on the ball."* In other words, apply rudder pressure to the side toward which the ball has moved. In figure 2-11, with the ball to the left of center, left rudder pressure should be applied.

STRAIGHT-AND-LEVEL FLIGHT BY INSTRUMENT REFERENCE

After practicing straight-and-level flight using visual references, the instructor will ask the student to put on a vision restrictor (IFR hood). With the hood on, all outside visual references are obscured.

The student then will practice straight-and-level flight using instrument references only. The aircraft will respond and react just as it did when it was flown using visual references. During flight by instrument reference, "attitude flying" will still be practiced utilizing the attitude indicator, airspeed indicator, vertical

velocity indicator, and altimeter for pitch control, and the turn coordinator and heading indicator for control of heading. (See Fig. 2-12.)

Flight instruments are scientifically arranged, in most modern aircraft, in a basic T configuration with the attitude indicator as the center of the T and supporting instruments beside and below to ease *scanning*. This arrangement is shown in figure 2-13.

Fig. 2-13. Basic "T" Instrument Arrangement

Once the aircraft is placed in straight-and-level flight, the student should adjust the miniature aircraft in the attitude indicator so the top of the aircraft is precisely in line with the top of the artificial horizon bar. This setting should be as viewed from the pilot's normal sitting position. If needed, the miniature airplane can be adjusted up or down to align with the horizon bar. The adjustment is made with the knob provided on the instrument.

The *attitude indicator* supplies the pilot with *both* wingtip position and nose position information. When using the attitude indicator in combination with the airspeed indicator and the altimeter for support, the student should be able to detect even minute changes in pitch and bank.

The horizon line is designed to remain parallel to the earth's horizon and is, in effect, a substitute or *artificial horizon*. If a wing dips or the nose position changes, the artificial horizon will move.

PITCH CONTROL

HEADING CONTROL

Fig. 2-12. Flight Reference Instruments

The attitude indicator shown in figure 2-14 indicates that the left wing is slightly down and the nose slightly up.

Deviation from desired attitudes and headings is caused by the same forces when flying by instrument references as when using visual references, and corrections are made in the same way. When it is noted that the attitude and heading have changed, or are changing, the pilot should first stop the turn by returning to the straight-and-level attitude. From the attitude indicator, the pilot should then refer to the other instruments (altimeter, airspeed indicator, and heading indicator) to determine the amount of correction needed to return to the orig-

inal heading and altitude. Then, he should make small attitude corrections to return to the desired readings.

When flying by instrument references, it is possible to experience *vertigo* (some feelings derived from the "seat of the pants" that are not in accord with the actual attitude and heading of the aircraft). These feelings can cause a new student to make an incorrect control movement; therefore, it is extremely important that the student believe *only* what his instruments tell him and not respond by feel alone.

MINIMUM STANDARDS OF PERFORMANCE FOR STRAIGHT-AND-LEVEL FLIGHT

The minimum acceptable standards for straight-and-level flight are recognition of proper attitude using either visual or instrument references and the ability to make prompt corrections. Altitude should be maintained *within 100 feet* of assigned altitudes, airspeed *within five knots* of the desired airspeed, and heading *within 10°* of the assigned heading.

The new student is not expected to perform to these standards on the first flight; however, these are pilot certification goals and, on each flight the student should strive to obtain performance that meets ever-narrowing tolerances.

Fig. 2-14. Attitude Indicator Showing Left Bank And Slightly Nose-High Attitude

SECTION B — CLIMBS AND DESCENTS

CLIMBS

The objectives when practicing climbs are to obtain proficiency in establishing the proper climb attitude, to apply the appropriate control pressures, and to correctly trim the aircraft in order to maintain the climb attitude. During practice climbs, emphasis is placed on learning the relationship between attitude and power, climb speed and climb performance.

ENTERING THE CLIMB

From straight-and-level flight, a climb is entered by increasing back pressure on the control wheel. The back pressure will raise the nose position of the aircraft. The aircraft nose position or pitch attitude should continue to rise smoothly until the desired climb-pitch attitude is established. The pitch attitude, as represented by the nose position and the attitude indicator, should resemble the indications shown in figure 2-15.

As the climb attitude is established, the airspeed will gradually slow and stabilize on or near the desired climb speed. As the desired climb speed is approached, the pilot should smoothly add power to establish the recommended climb power setting. The climb attitude plus the climb power setting will determine the aircraft's climb performance.

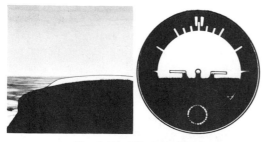

Fig. 2-15. Climb Attitude

Since the airspeed has changed, the pilot will find that he is now continually hold-

ing back pressure on the control wheel in order to maintain the climb attitude. Therefore, he should make a trim tab adjustment to relieve control pressures.

The position of the wingtips and the angle which they make with the horizon can be used to assist in establishing the proper climb attitude in the same way that these indications were used to establish level flight attitudes. For example, in the left half of figure 2-16, the left wingtip is seen as it appears when the airplane is flying level for both high- and low-wing aircraft. By way of contrast, the same wingtip is shown in the right half of figure 2-16 when the aircraft is in the climb attitude.

It will be found that the airspeed indicator serves as a secondary instrument in determining if the proper climb airspeed is being maintained. If the pilot observes that the airspeed is either lower or higher

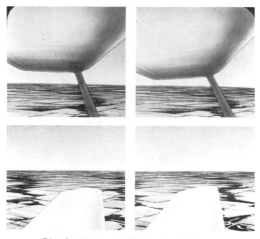

Fig. 2-16. Level Flight And Climb Wing Tip Attitudes

than desired, he should use the available attitude references to adjust the aircraft's nose position and then retrim slightly when the new attitude produces the desired climb speed.

LEFT-TURNING AND ROLLING TENDENCIES DURING CLIMBS

As the climb is established, the pilot will notice that the aircraft may tend to roll to the left and that the ball of the turn coordinator tends to move off center to the right. This indicates that right rudder pressure is required to keep the aircraft streamlined and properly coordinated and to return the ball to the center of the inclinometer.

The left-rolling tendency is caused by a combination of forces such as P-factor, torque, and spiralling slipstream. These effects are most pronounced at high power settings and low airspeeds, such as experienced during climbs.

The manufacturer of the airplane "rigs" the airplane to compensate for the left-rolling tendency for *cruising flight*. However, at other power settings or speeds, it is necessary that the pilot apply some aileron and rudder pressures to keep the aircraft from rolling and skidding.

Another method used to counteract the left-rolling and left-turning tendency on more powerful aircraft is shown in figure 2-17. *Aileron and rudder trim tabs* permit the pilot to compensate for turning effects such as those produced during climbs.

MAINTAINING THE CLIMB

In order to monitor whether he is achieving the proper climb airspeed and heading, the pilot should refer to the support instruments (airspeed indicator, altimeter, and ball of the turn coordinator). If the indications are not as desired, small adjustments in attitude should be made

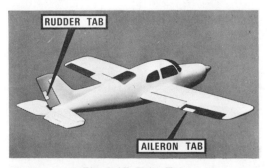

Fig. 2-17. Trim Tabs

by reference to the nose and wingtip positions, or attitude indicator. Then, the aircraft should be permitted to stabilize in the new attitude and control pressures removed by trimming.

CLIMB SPEEDS

In the early phases of flight training, an instructor will designate one airspeed as the *normal climb speed* and this speed will be used during climb to practice altitude. However, before the training program has been concluded, three additional climb speeds will normally be specified and practiced. Each of the following speeds is used to achieve a different aircraft performance capability.

1. The *cruising climb speed* is used to achieve a satisfactory groundspeed while climbing to cruising altitude during cross-country flight. This speed is usually slightly higher than normal climb speed and will provide adequate engine cooling. The pilot can determine the cruising climb speed by referring to the aircraft owner's manual.

2. Also listed in the owner's manual is the *best rate-of-climb speed*. This airspeed is lower than the cruise climb speed and provides the *most gain in altitude per minute*; therefore, it is the speed utilized to get the aircraft to the desired altitude in the *shortest amount of time*, as shown in figure 2-18. The pilot should recognize that he will not gain altitude faster at any speed higher or lower than the designated best rate-of-climb airspeed.

3. Another airspeed listed in the owner's manual is called the *best angle-of-climb speed* and is generally the lowest of the specified climb speeds. This speed results in a steeper angle of climb and is

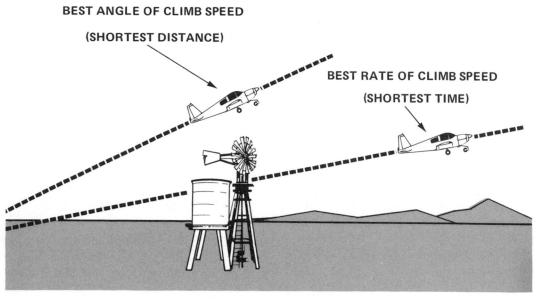

BEST ANGLE OF CLIMB SPEED

(SHORTEST DISTANCE)

BEST RATE OF CLIMB SPEED

(SHORTEST TIME)

Fig. 2-18. Climb Speeds

used to clear obstacles in the take-off path, such as trees or power-lines at the end of the runway. Fig. 2-18 compares the climb profile of best angle and best rate-of-climb speed and illustrates that the *best angle-of-climb speed results in the greatest altitude gain in the shortest distance.*

LEVEL-OFF FROM A CLIMB

To return to straight-and-level flight from a climb, it is necessary to begin the transition to level flight *before* reaching the desired altitude. As when driving a car, the pilot must start "slowing up" before reaching the point where he wants to stop. In the airplane the pilot must begin slowing the rate of climb before reaching the desired altitude.

Approximately 20 feet below the desired altitude, the nose should be smoothly lowered to the level flight attitude and attention given to holding this attitude by visual or instrument references. (See

Fig. 2-19.) Climb power is maintained until *reaching cruise speed;* then, power is reduced to the cruise setting and control pressures required to hold the aircraft in level flight attitude are trimmed off.

ACCEPTABLE CLIMB PERFORMANCE

Acceptable performance for climbs includes prompt recognition of the proper climb attitude, the ability to maintain airspeed within five knots of the desired speed, and the control of power within

Fig. 2-19. Leveling Off From A Climb

50 r.p.m. or one inch of mercury on the manifold pressure gauge. Also, the ability to effect the rudder and aileron coordination needed to counteract the cumulative left-turning and left-rolling tendencies should be exhibited. Furthermore, if the aircraft's nose blocks the view of other aircraft in the vicinity of the climb, the pilot should make "S-turns" to constantly clear the area into which he is climbing.

DESCENTS

The objectives of practicing descents are to lose altitude without excessive build-up of speed, to convert altitude into as much distance as is possible to achieve without power, to control the rate of descent with power and attitude, and to learn the different control pressures necessary for reduced power descending flight.

Descents are practiced at a speed near that used for approaches to landings, or the *best glide* speed. Initially, the instructor will specify a descent speed from the listings in the aircraft owner's manual.

Most light training aircraft have power-off glide ratios of approximately 10:1. Stated another way, this means the aircraft will go forward 100 feet for every 10 feet of altitude that it loses. Descents *with power* result in glide ratios *higher* than 10:1, the ratio increasing as the amount of power used during the descent increases.

ESTABLISHING A DESCENT AT APPROACH SPEEDS

Many aircraft manufacturers recommend that carburetor heat be used anytime there are prolonged periods of low power settings; so the first step in establishing a descent usually is to apply carburetor heat. A flight instructor or the owner's handbook should be consulted for the recommended procedures for the training aircraft used.

To begin the descent, power is reduced to a predetermined setting or to an idle. As the power is reduced, back pressure is gradually applied to the control wheel to reduce the airspeed. This means that the nose attitude of the aircraft will gradually rise from the normal cruise straight-and-level position.

When the desired descent airspeed is reached, the nose attitude is lowered to the descent attitude to hold this airspeed. After the proper descent attitude is established, the trim control should be used to remove elevator or stabilator back pressure. Many beginning pilots are surprised to find that this pitch attitude may be *nearly the same* as the attitude used for straight-and-level cruise flight.

The reason for the level descent attitude may be easily understood by referring to the diagram in figure 2-20, which illustrates the aircraft's flight path and angle of attack in a descent. In order to produce lift at slower airspeeds, the wing must be at a higher angle of attack. During a descent, the attitude required for the proper angle of attack is near that required for level flight at cruise airspeeds.

During the descent, the student will observe some right rolling and yawing effects *opposite* to those encountered during a climb. Since the tendency to roll and yaw to the right is not as noticeable in descents as it is in climbs, small amounts of left aileron and rudder pressures are required for proper coordination. Because the airflow over the controls is less rapid than at cruise airspeed and there is little propeller slipstream over the empennage, the controls will tend to feel soft or mushy and less positive in response. The pilot can learn to use this "feel" as he develops his skills to help determine when the aircraft is flying slowly. The feel developed in practicing descents will find an important application during approach and landing practice.

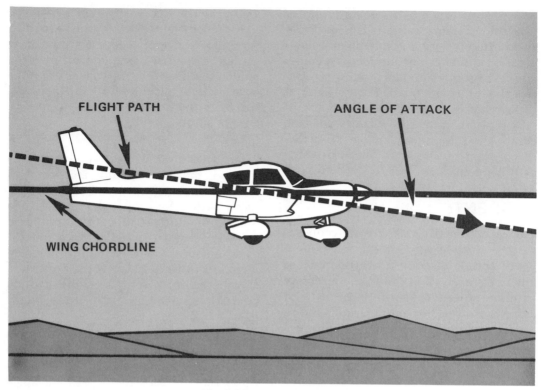

FLIGHT PATH

ANGLE OF ATTACK

WING CHORDLINE

Fig. 2-20. Descent Attitude

MAINTAINING THE DESCENT

The support instruments (airspeed indicator, altimeter, heading indicator, turn coordinator, and vertical velocity indicator) are used during the descent as in other maneuvers. After the pilot has established the desired descent attitude and has the aircraft properly trimmed, he should refer to the support instruments to affirm that the airspeed, heading, and rate of descent are as desired. If an adjustment is required, he should return to the attitude clues, either visual or instrument, make an attitude adjustment, permit the attitude to stabilize, and then refer to the appropriate instruments to *confirm* that he is descending as desired. For example, if back pressure is required, the desired nose *attitude is established* with elevator control pressure. Then the *pressures are relieved* with the trim tab. If the pilot tries to establish the attitude using the trim tab *rather* than with control pressures, he will usually overcontrol the aircraft and make an erratic descent.

CONTROLLING THE RATE OF DESCENT

USE OF POWER

The rate of descent can be controlled with power. If the pilot wishes to decrease the rate of descent, he should add power; to increase the rate of descent, he should reduce power. (See Fig. 2-21.) As power is added, the nose attitude should be held slightly higher if the pilot desires to maintain a constant airspeed. On the other hand, a power reduction must be accompanied by lowering the pitch attitude slightly in order to maintain a given airspeed. When power is added, propeller slipstream over the elevator or stabilator is increased so the pilot will find that a small trim tab adjustment is necessary.

The rate of descent can be monitored by reference to the vertical velocity indicator once the descent is well established and the instrument has settled to an accurate indication. The descent airspeed is controlled by *changing the pitch attitude* in

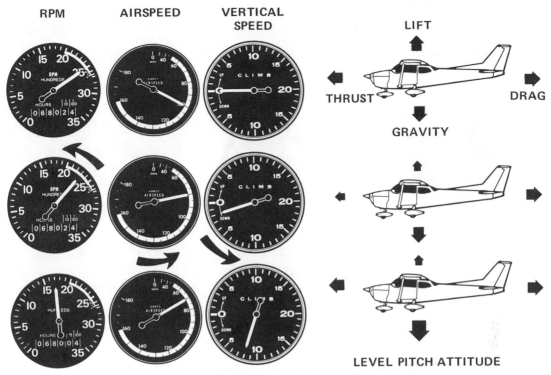

RPM   AIRSPEED   VERTICAL
SPEED

LIFT

THRUST   DRAG

GRAVITY

LEVEL PITCH ATTITUDE

Fig. 2-21. Rate Of Descent Can Be Controlled By Power

the same manner as when performing climbs or straight-and-level flight.

USE OF FLAPS

The *angle* of descent can be *increased* by the use of flaps. As flap deflection increases, it will be necessary for the pilot to lower the pitch attitude, as shown in figure 2-22, if he desires to maintain a constant airspeed. Moderate trim tab adjustments may also be necessary.

Descents with flaps up are made when the pilot desires the maximum range or distance. When rapid dissipation of altitude is desired, a full-flap, power-off glide should be used. (See Fig. 2-23.)

LEVEL-OFF FROM A DESCENT

To return to straight-and-level flight, it is again necessary for the pilot to begin the transition to level flight before reaching the desired altitude. Approximately 50 to 100 feet above the desired altitude, the pilot should adjust the nose position to level flight and simultaneously add

power to the cruise setting. The pilot should refer to wing and nose positions in order to maintain the proper attitude throughout the transition from descent to straight-and-level flight. Since the addition of power and the increase in airspeed will produce a *moderate* tendency to *pitch upward*, the trim tab must be adjusted to relieve forward control pressures as straight-and-level flight is attained.

ACCEPTABLE DESCENT PERFORMANCE

The criteria used to evaluate acceptable descent performance is prompt recognition of the descent attitude, airspeed maintained within five knots of the desired speed, power within 50 r.p.m. or one inch of mercury on the manifold pressure gauge, and the application of coordinated rudder and aileron pressures to counteract any rolling or turning tendencies. Also, the heading should be maintained within 10° of an assigned heading.

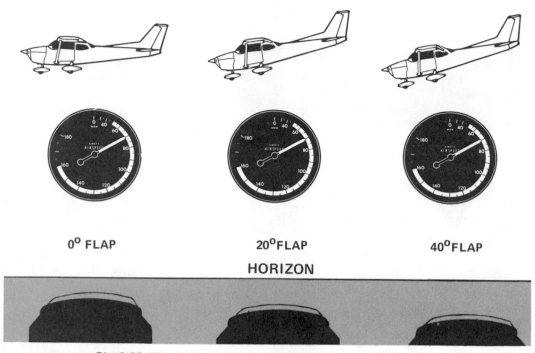

0° FLAP 20°FLAP 40°FLAP

HORIZON

Fig. 2-22. Nose Attitude Changes With Application Of Wing Flaps

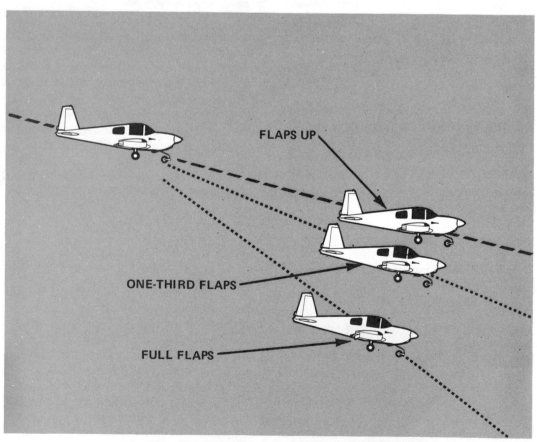

FLAPS UP

ONE-THIRD FLAPS

FULL FLAPS

Fig. 2-23. Wing Flaps Effect Rate Of Descent And Range

SECTION C – TURNS

Turns are described in general terms by the number of degrees of bank necessary to produce the turn. A *medium bank turn* is one in which the bank is approximately 30°. The medium bank is used for training in preference to shallower banks since most training aircraft have a tendency to *return* to level flight due to their inherent stability.

WHY THE AIRCRAFT TURNS

The objectives of turns are to change the direction of the aircraft's flight path and develop proficiency in control coordination. Before practicing turns, it is important to review *what makes the aircraft turn*. Turns are made by directing a portion of the lift force of the wings to one side or the other. As shown in figure 2-24, lift is equal to the force of gravity when the aircraft is in level flight. However, when the aircraft turns, the lift force is tilted out of alignment with gravity and produces a force to turn the

aircraft. (See Fig. 2-25.) The lift force in a turn can be subdivided and represented as two forces or *components*, one acting vertically and one acting horizontally. The *horizontal* component is the force that makes the aircraft *turn*, while the *vertical* component is the force that *overcomes gravity*.

An analysis of the forces acting on an airplane in a turn is shown in figure 2-26. When the aircraft is in a level turn, the vertical component of lift is opposed by an equal force acting in the opposite direction, gravity. The horizontal component of lift, sometimes called *the turning* force, is opposed by an equal force

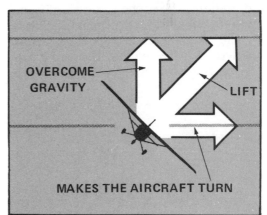

Fig. 2-25. Lift Components In A Turn

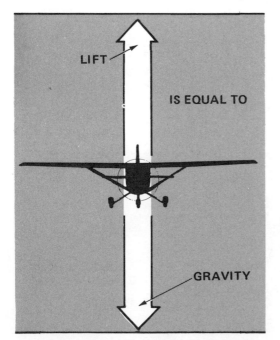

Fig. 2-24. Lift Equals Gravity In Level Flight

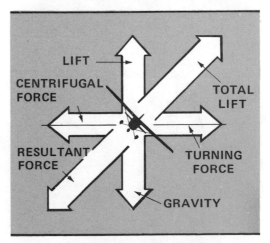

Fig. 2-26. Forces Acting On
An Aircraft In A Turn

acting in the opposite direction, *centrifugal* force. The total lift force created by the wings is opposed by the combination of centrifugal force and gravity acting in the opposite direction to total lift. This force is termed the *resultant* force.

Since the resultant force is a combination of centrifugal force and gravity, the pilot, in a properly executed turn, feels *pressed into the seat* with a force somewhat greater than in straight-and-level flight. The pilot should experience no forces that tend to make him *lean* into or away from the turn.

Figure 2-27 illustrates another important concept. The *steeper* the bank used to make the level turn, the *greater* the total lift required to support the aircraft and produce the turn. This principle is shown by the comparative lengths of the arrows labeled "T." Notice that the arrow illustrating the turning force "T" in a shal-

low-banked turn is considerably shorter and of less magnitude than arrow "T" for the aircraft making a steep-banked turn. Also, the centrifugal force represented as "C" is considerably greater in a steep-banked turn than in a shallow-banked turn. Figure 2-27 *also* illustrates that the *resultant force pressing the pilot into the seat increases as the angle of bank increases.*

In review, when the aircraft is in level flight, the total lift force is equal to gravity, as shown on the left in figure 2-28. When the aircraft is banked, the total lift force is diverted. While the total lift is still equal to gravity, the vertical component is *insufficient* to counteract gravity and the aircraft will *descend*. Therefore, in order to maintain altitude during a turn, the total lift must be increased until the vertical component is equal to gravity. This is accomplished by *increasing the angle of attack* with back pressure on the control wheel.

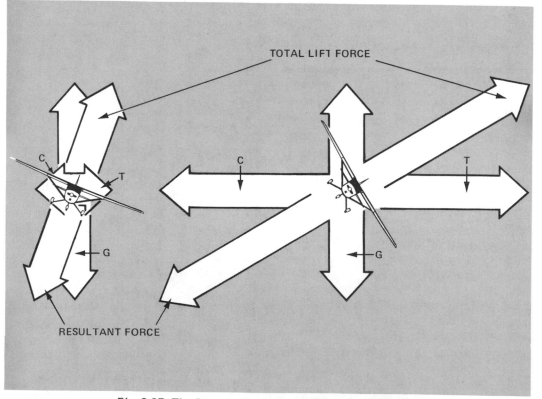

Fig. 2-27. The Steeper The Bank, The Greater The Forces

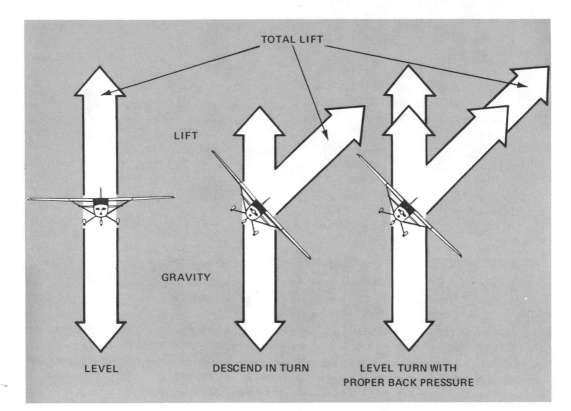

Fig. 2-28. Back Pressure On The Control Wheel Required In A Turn

MAKING THE AIRCRAFT TURN

AILERON CONTROL PRESSURES

To roll into a bank, aileron control pressure is applied to the control wheel in the direction toward which the pilot desires to turn. When executing a *left turn*, the control pressures will place the *left aileron up* and the *right aileron down*. As shown in figure 2-29, this action causes the right wing to produce more lift than the left wing, thereby making the aircraft roll to the left. How *fast* the airplane rolls depends on how much aileron control pressure is applied. How *far* the airplane rolls (the steepness of the bank) depends on how *long* the ailerons are deflected, since the airplane will continue to roll as long as the ailerons are deflected.

When the airplane reaches the desired angle of bank, the ailerons are *neutral-*

ized, since after the bank is established, the turn will *continue* until the pilot moves the controls in the opposite direction. (See Fig. 2-30.)

ELEVATOR CONTROL PRESSURES

Since the pitch attitude must be increased slightly to increase the total lift when performing a turn, the nose position used to maintain the desired altitude is *slightly higher* in a turn than in level flight. (See Fig. 2-31.) In a medium-

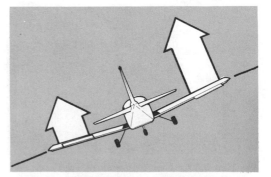

Fig. 2-29. Ailerons Cause Aircraft To Roll

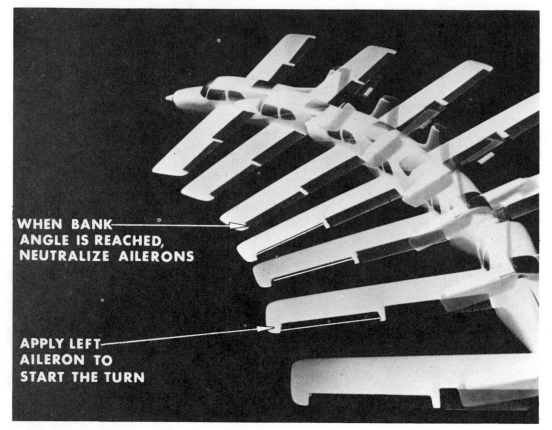

WHEN BANK
ANGLE IS REACHED,
NEUTRALIZE AILERONS

APPLY LEFT
AILERON TO
START THE TURN

Fig. 2-30. Aileron Movement In A Left Turn

banked turn, the required pitch attitude adjustment is slight and almost imperceptible. However, the steeper banks practiced later require a *much higher* nose position adjustment, as illustrated in figure 2-32.

COORDINATION OF CONTROLS

When discussing turning techniques, pilots use the term *coordination*. This means the simultaneous application of rudder pressure each time aileron pressure is applied and application of correct elevator back pressure at the proper time.

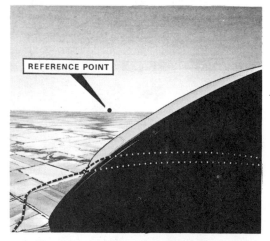

Fig. 2-31. Nose Position And Reference Point In Medium-Banked Turn

Fig. 2-32. Nose Position And Reference Point In Steep-Banked Turn

The use of rudder pressure is made necessary by an unbalanced aerodynamic condition created when the ailerons are deflected. When the aileron is deflected downward, as shown on the right wing in figure 2-33, there is an increase in lift on that wing but the drag force also increases. The increase in *drag force is greater on the right wing* than the increase in drag from the upward deflected left aileron. This is because an *increase in lift* is always *accompanied by an increase in drag* and a *decrease in lift* is accompanied by a *decrease in drag*.

The increased drag on the right wing tends to make the nose move or *yaw* in a *direction opposite* to the intended turn. The production of a yaw opposite to the intended direction of turn is known as the *adverse yaw effect*. Therefore, it is necessary to apply just enough rudder pressure *in the direction of the turn* (to the right in this example) to counteract the yaw. (See Fig. 2-34.) This rudder pressure is maintained *as long as* the ailerons are deflected; and when the ailerons are neutralized, the rudders are *also* neutralized. The actual amount of rudder pressure, compared to aileron pressure to counteract adverse yaw, varies with different types of airplanes and is determined and perfected by practice.

The ball of the turn coordinator will assist the pilot in determining how much pressure is necessary to counteract adverse yaw. For a coordinated turn, the ball should remain in the center of the inclinometer. If the ball is not in the center, as demonstrated in figure 2-35, the pilot should apply rudder pressure on the side to which the ball has rolled; in other words, "step on the ball."

While the turn is in progress, the pilot should occasionally check the support instruments (altimeter, vertical velocity indicator, and airspeed indicator) to determine if corrections may be necessary. If corrections are required, the pilot should make attitude adjustments using the

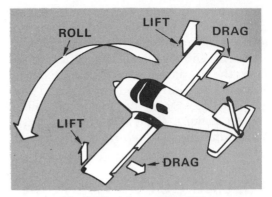

Fig. 2-33. Adverse Yaw Effect

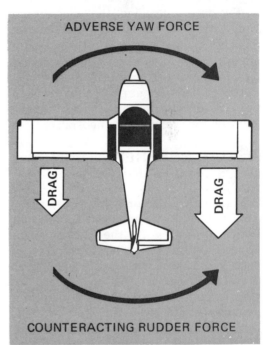

Fig. 2-34. Rudder Is Used To Counteract Adverse Yaw

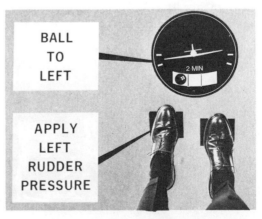

Fig. 2-35. "Step On The Ball"

"two-step" method. For example, if the nose is low, the airspeed high, and the altitude decreasing, a pitch attitude adjustment should be made sufficient to *hold* the altitude. Once the airspeed has returned to the desired reading and the altitude loss has stopped, the pilot should make a slight nose-up pitch attitude adjustment to *return* to the desired altitude.

ROLLING OUT OF THE TURN

Approximately 5° to 10° *before* reaching the desired heading, the pilot should apply coordinated aileron and rudder pressures to roll out of the turn. Simultaneously, the pilot must begin *releasing* back pressure held on the control wheel so that when the airplane has reached the wings-level position, aileron, rudder, and elevator pressures are neutralized.

VISUAL REFERENCES IN THE TURN

The pilot making turns by visual reference will apply coordinated aileron and rudder pressures in the direction he wishes to turn. As the pilot looks to the front, the nose of the aircraft will appear to roll with respect to the horizon. To assist in coordinating aileron and rudder pressure, again project a line of sight from the pilot's eye through the windshield (through the "spot" if one is used) to a point on the horizon. Figure 2-36 shows that the point at which the line of sight meets the horizon should begin *moving in the same direction* and at the

Fig. 2-36. Visual Reference Point Moves With Bank Angle

same rate as the rate and direction of bank establishment. The angle at which the cowling, instrument panel, and door posts meet the horizon will assist in determining when the proper angle of bank has been attained. Also, in the 15° shallow-banked turn, the left wingtips on typical low-wing or high-wing airplanes look like those shown in figure 2-37.

It will be found that the nose position appears different when making left and right turns. This is because the pilot sits to the side of the centerline of the airplane. Both of the banks shown in figure 2-38 are approximately 15°. The one on the left looks nose-high and the one on the right looks nose-low. Therefore, the point *directly in front of the pilot's line of sight* (again, the "spot" placed on the windshield, if one is used) is the point that should be used for reference during the turn. If the pilot attempts to sight over the center of the nose of the aircraft, he will introduce an error in establishing the proper pitch attitude. The steeper the angle of bank, the more pronounced this illusion will be.

TURNS BY INSTRUMENT REFERENCE

After some practice making turns by visual references, the student will perform turns using instrument references. Control pressures are applied in the same manner as when using visual references and the support instruments are interpreted similarly.

PERFORMING THE TURN

The turn by instrument reference is established in the same manner as when using visual references except the references have changed from the wingtips and nose *outside* the aircraft to the wing and nose positions as represented on the *attitude indicator*. To establish a 15° banked turn, the airplane is banked until the 15° mark is aligned with the bank indice at the top of the attitude indi-

Fig. 2-37. Wing And Nose Attitudes in a 15° Turn

*Fig. 2-38. Nose Position Appears Different
In Left And Right Turns*

cator, as shown in figure 2-39. Then, the aileron and rudder pressures are neutralized and the pitch attitude is adjusted slightly upward with additional back pressure on the control wheel.

Corrections are made in the same manner as when using visual references; however, the pilot adjusts the attitude by using the attitude indicator and scans the support instrument group to determine that the turn is progressing as desired. One advantage of performing turns by instrument reference as compared to visual reference is that the nose position on the attitude indicator looks the same during both a left and a right turn, as illustrated in figure 2-40.

Recovery from the turn is performed by applying coordinated aileron and rudder pressure, releasing the back pressure, and reestablishing straight-and-level flight by reference to the attitude indicator.

STANDARD-RATE TURNS

The angle of bank frequently used in making turns by instrument reference is one that results in a *standard-rate turn.* The definition of a standard-rate turn is one that produces a turn rate of *three degrees per second.* The pilot uses the turn coordinator to indicate the direction and rate of turn. When the wing is aligned with the indice, a standard-rate turn of 180° in one minute is indicated. (See Fig. 2-41.) If the wing of the little airplane is half way between the level flight position and the standard-rate position, a *half standard-rate turn* or turn of 180° in *two minutes* will be performed.

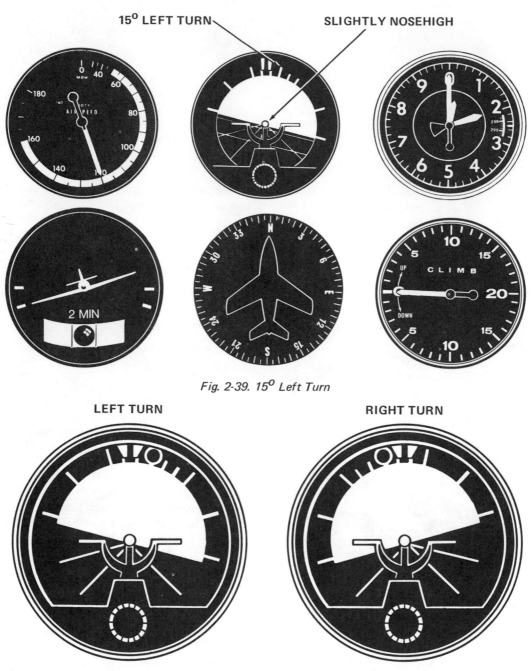

15° LEFT TURN   SLIGHTLY NOSEHIGH

Fig. 2-39. 15° Left Turn

LEFT TURN   RIGHT TURN

Fig. 2-40. Left And Right Instrument Turns Have Same Nose Position

ANGLE OF BANK REQUIRED FOR STANDARD-RATE TURN

When an aircraft is performing a standard-rate turn, it will turn 3° in 1 second, 30° in 10 seconds, 90° in 30 seconds, or 180° in 60 seconds. The *angle of bank* necessary to produce the standard rate turn is strictly a *function of the airspeed* — the *greater* the airspeed, the larger the angle of bank required to maintain a standard-rate turn.

Since most training aircraft maintain cruise true airspeeds of approximately 120 miles per hour, the angle of bank required for a standard-rate turn is ap-

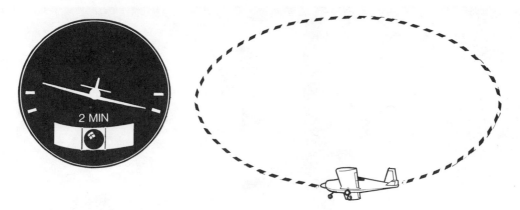

Fig. 2-41. Standard-Rate Turn (3 Degrees Per Second)

FLIGHT CONDITION	SPEED	ANGLE OF BANK
CRUISE	120 MPH	17°
CLIMB	80 MPh	12°
DESCENT	75MPH	11°

Fig. 2-42. Standard-Rate Turn Angle Of Bank Vs. Airspeed — Typical Training Aircraft

proximately 17°. Since climbing and descending turns are normally performed at airspeeds below 100 miles per hour, a standard-rate turn would require a bank angle of less than 14°. The bank angles for standard-rate turns during cruise, climb, and descent for the typical training aircraft are shown in figure 2-42.

TIMED TURNS

By utilizing a standard rate, the pilot can determine the *amount of time* required to make a turn by instrument reference *prior to initiating the turn*. He should determine the number of degrees to be turned and *divide by three*. Then, noting the time according to the second hand on the aircraft clock, he should roll into the turn using the proper angle of bank and roll out to level flight when the predetermined amount of time has elapsed. For example, to make a 180° turn, the pilot simply divides 180 by 3 and finds that it will take 60 seconds to perform this turn. (See Fig. 2-43.)

ACCEPTABLE PERFORMANCE IN TURNS

Acceptable performance in turns is characterized by prompt recognition of proper aircraft attitude using either visual or instrument references. If the turn is properly coordinated, the ball should not move more than one-half the ball diameter out of the center of the inclinometer. Corrections must be prompt and proper. Bank should be maintained within five degrees of the desired bank, altitude should be held within 100 feet of an assigned altitude, and the recovery from the turn made within 10° of an assigned heading.

COMBINATIONS OF FUNDAMENTAL MANEUVERS

Four maneuvers — *straight-and-level flight, turns, climbs*, and *descents* — make up the fundamental elements of flight techniques. All other flight maneuvers consist of either one or a combination of these fundamental maneuvers. If the student develops a thorough understanding of these basic elements and obtains effective and precise control of the aircraft when performing each maneuver, he will develop a high level of proficiency as he continues his flight training.

After practicing the four fundamental maneuvers individually, the student will

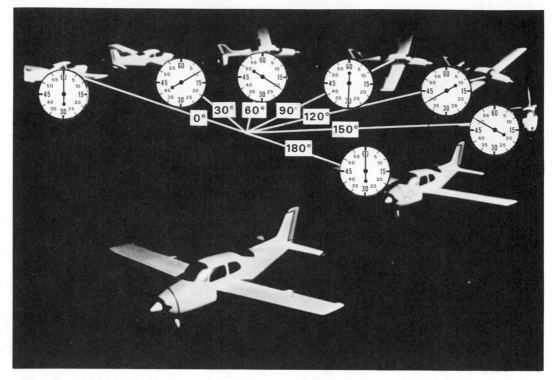

Fig. 2-43. 60 Seconds Required For A 180-Degree Standard-Rate Turn

practice combinations of them. Climbing and descending turns to a predetermined heading and altitude will be introduced as soon as the student has obtained proficiency in climbs, descents, and turns. These combination maneuvers are practiced using both visual and instrument references.

CLIMBING TURNS

The objective of practicing climbing turns is to smoothly combine the techniques of climbs with those used in performing turns. The climbing turn is used in reaching traffic pattern altitude after takeoff and when climbing to a selected cruise altitude and heading.

To perform a climbing turn, the student should establish the climb as previously discussed. When climb power and attitude are set, he should then roll to the desired bank angle. This is a two-step procedure initially; however, as the student gains experience and proficiency, he will perfect the maneuver by *simul-*

taneously establishing the climb attitude and the proper bank.

When performing climbing turns, it is recommended that the airspeed *remain the same* as in the straight climb. Thus, the pilot accepts a reduction in the rate of climb when performing climbing turns at a constant airspeed. Since a further increase in the angle of bank will divert more of the total lift to make the aircraft turn and cause a further reduction in rate of climb, climbing turns are generally performed at *shallow* bank angles.

It is unlikely that the student will arrive at the desired heading and altitude at the same time; therefore, if the desired heading is reached first, the wings should be leveled and the climb maintained until the desired altitude is reached. On the other hand, if the altitude is reached first, the nose should be lowered to a level flight attitude and the turn continued to the desired heading. If both the desired heading and altitude are reached at the same time, the pilot can perform the level-off procedure simultaneously.

ACCEPTABLE PERFORMANCE IN CLIMBING TURNS

When performing climbing turns, the student should demonstrate the ability to establish the proper bank and pitch attitudes, be able to maintain airspeed within five knots of the desired speed, and he should be able to recover from the turn to within 10° of an assigned heading, and level off within 100 feet of an assigned altitude. The student should also demonstrate the prompt and effective use of trim to relieve control pressures.

DESCENDING TURNS

Descending turns to preselected headings and altitudes combine the procedures for a straight descent with those used in turns. The pilot should enter the descent using either visual or instrument references in the same manner as previously outlined for the straight descent. Once a descent attitude has been established, he should roll to the desired angle of bank. As with climbing turns, the initial procedure is performed in a two-step manner; however, as the student gains profi-

ciency, the descent attitude and the bank are established simultaneously. As in any maneuver, control pressures required to maintain the selected attitude should be trimmed off.

When using visual references, the nose will look lower in the right turn than in the left turn although both turns are performed at the same airspeed and rate of descent. (See Fig. 2-44.) When performing descending turns by instrument references, the attitude will look the same in both left and right turns, as illustrated in figure 2-45. In most training aircraft, the attitude shown will produce a standard rate of turn with approximately a 500-foot-per-minute rate of descent.

Power is used to control the rate of descent. The pilot should make the initial power setting for the desired rate of descent and allow the pitch attitude and the rate of descent to stabilize. If a higher rate of descent is desired, power should be reduced. In contrast, the pilot must add power if the rate of descent is higher than desired.

Fig. 2-44. Nose Attitudes For Descending Turns

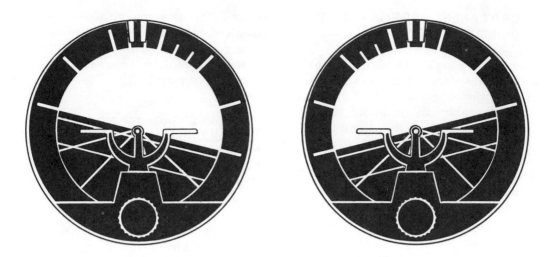

Fig. 2-45. Instrument Attitudes For Descending Turns

It is recommended that the same airspeed be maintained in descending turns as in straight descents. Because the vertical lift component is less when the aircraft banks, the pilot will find that the rate of descent is *higher* in a descending turn than in a straight descent with comparable power settings. Compensation may be made with a slight addition of power.

It is unlikely that the student will arrive at the desired altitude and heading at the same time; therefore, approximately 50 to 100 feet before reaching the desired altitude, the student should adjust the nose slightly to achieve level flight attitude (if not already attained) and increase the power setting to cruise power. Approximately 10° before reaching the desired heading, the rollout to a wings-level position should be initiated. If the heading and altitude are reached at the same time, the recovery procedures should be made simultaneously.

ACCEPTABLE PERFORMANCE IN DESCENDING TURNS

Acceptable performance in descending turns is demonstrated when the airspeed is maintained within 10 knots of the desired speed, the bank angle held within 10° of the desired angle, and the turn recovery made within 10° of an assigned

heading. The level-off should also be completed within 100 feet of a preselected altitude. In addition, the student should demonstrate that he recognizes and can establish the correct attitude for descending turns and show no tendency to allow the nose to drop excessively when in the turn. The trim tab should be used as required to relieve control pressures.

COORDINATION EXERCISES

To develop and maintain proficiency in the coordination of the flight controls, certain exercises not normally used in everyday flight are utilized. One of the most common of these exercises, called *dutch rolls*, consists of maintaining straight-and-level flight while rolling back and forth from right to left banks without stopping at the wings-level position. (See Fig. 2-46.) The nose is held on a point or heading and is not allowed to rise or fall. The degree of bank at which the rolling is reversed may be shallow or steep or may vary from shallow to steep as the maneuver is performed.

Although dutch rolls require coordination of the rudder and ailerons in *exactly the opposite manner* to that used in normal turns, they are an excellent aid in smoothing control usage and in learning control reactions. Also, as steeper banks are used, a high degree of elevator coordination is required to keep the nose from pitching up or down in relationship to the horizon.

Another coordination exercise consists of rolling from one medium turn directly into another in the opposite direction and reversing the turn after a predetermined number of degrees of turn. For example, the turn can be reversed after 90° of turn and the aircraft turned through an arc of 180°; then, the direction reversed again. As proficiency is attained, the number of degrees of turn can be reduced. For example, turning 20° to 30° and then rolling into a turn in the opposite direction for 20° to 30°

is a valuable assist to the pilot in perfecting his coordination techniques.

An advanced coordination exercise is illustrated in figure 2-47. The student begins the exercise with a shallow 15° bank, turns right for 90°, and then turns left for 90°. The exercise is continued by rolling into a 30° banked turn, continuing this turn through 180°, and then reversing the bank and turning another 180°. The exercise is terminated by flying a 360 turn with a 45 bank followed by a similar banked turn of 360 in the opposite direction. Upon completion, the pilot should be on the same heading as when he began. This exercise is considered a good "warm-up" exercise before doing advanced maneuvers.

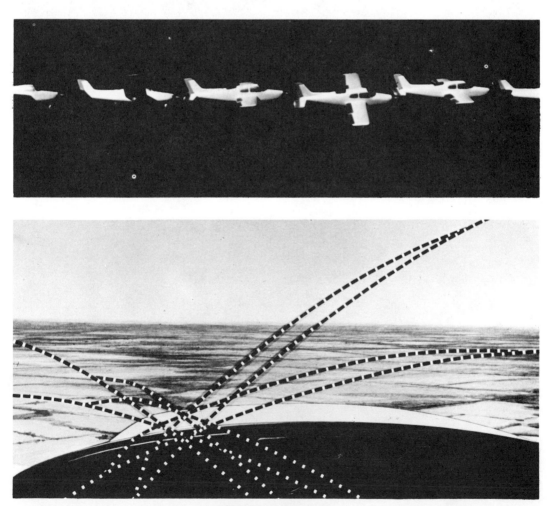

Fig. 2-46. Dutch Rolls

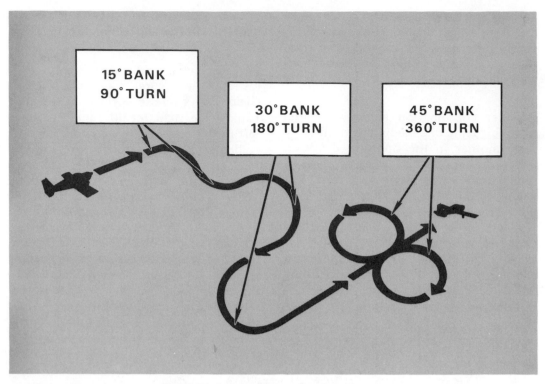

Fig. 2-47. Advanced Coordination Exercise

The student may be asked, as he becomes more adept, to trace an imaginary circle or square on the horizon with the extended centerline of the airplane's fuselage. The instructor may ask the student to perform any one or a number of the exercises as the student's flight technique requires.

CHAPTER 3

**TAKEOFFS
LANDINGS
AND SLIPS**

SECTION A – NORMAL AND CROSSWIND TAKEOFFS

Takeoffs and landings can be described simply as the transition of the airplane between being a ground-operating machine and an airborne-operating machine. The pilot's objective is simply to get the aircraft safely from the ground to the air and back again to the ground. In general, there are four types of takeoffs and landings which are used: *normal, crosswind, shortfield, and soft field.* This section is devoted to the normal and crosswind takeoffs and landings.

PRETAKEOFF CHECK

The takeoff begins with the pretakeoff checklist. The pilot should refer to the owner's handbook for the manufacturer's recommended procedures. To that list, the instructor may wish to add some items. Whatever the list consists of, the point is, before taking off, the pilot must have a methodical way of checking all systems so he will have reasonable assurance and confidence that they are operating properly. Pilots are strongly urged to use a written checklist prior to

each takeoff. Figure 3-1 shows a typical pretakeoff checklist.

BEFORE TAKEOFF
(1.) Cabin Doors – Latched.
(2.) Seat – Check position and security. Check that seat belt and shoulder harness are fastened.
(3.) Flight Controls – Check for free and correct movement.
(4.) Elevator Trim Control – "TAKEOFF" setting.
(5.) Throttle – 1700 r.p.m.
(6.) Engine Instruments – Within green arc.
(7.) Suction Gauge – Check (4.6 to 5.4 inches of mercury). r.p.m.
(8.) Magnetos – Check (75 r.p.m. maximum differential between magnetos).
(9.) Carburetor Heat – Check operation.
(10.) Flight Instruments and Radios – Set.
(11.) Emergency Locator Transmitter – Check.

Fig. 3-1. Typical Before Takeoff Checklist

The pretakeoff checklist outlines procedures for checking all systems, instruments, and radio gear. This should not be done while taxiing because it will distract the pilot's attention.

Before starting the runup checklist, the pilot should be sure that the airplane is positioned out of the way of other aircraft near the takeoff end of the runway. The nose should be pointed as nearly as possible into the wind to enhance engine cooling. Engine runups over loose gravel and sand should be avoided or damage to the propeller and other parts of the airplane may occur.

The engine normally will have warmed up enough before the pretakeoff check so that it will accelerate smoothly when it is "run up" to check the carburetor heat, ignition, and mixture. All engine instrument indications must be normal before takeoff.

During the engine runup, the pilot should divide his attention between the cabin and the area around the airplane. If the parking brake slips, or if he is not holding the toe brakes enough, the airplane could move forward while his attention is on the panel.

If any abnormal condition is observed during the runup, the airplane should be returned for maintenance, since even supposedly minor malfunctions could later affect the safety and efficiency of the flight. The checklist should be used for every runup; even after a very brief stop, all systems should be checked for proper operation.

A typical checklist is performed as follows:

1. Cabin doors — are checked, securely latched, and locked.

2. Seat — is adjusted to the proper position. This position must permit the pilot to obtain full deflection of the rudder in either direc-

tion. At the same time, a check should be made to insure the seat latches are engaged. Also, check that the seatbelt and shoulder harness are fastened and adjusted to a comfortable tension.

3. Flight controls — are checked to determine that they move freely and easily throughout their total travel. A brief wiggling of the controls is not sufficient; each control is moved through its range of travel until it hits the "limit stops." Also, the direction of travel of each surface is noted while the control is being moved.

4. Elevator trim control — is set to the TAKEOFF position.

5. Throttle — is set to the r.p.m. for the power check (usually 1,700 r.p.m. or as recommended in the owner's manual).

6. Engine instruments — are checked unhurriedly and should indicate operation in the green arcs.

7. Suction gauge — is checked for normal indication (usually 4.6 to 5.4 inches of mercury). A low reading usually indicates a dirty air filter. Unreliable gyro indications may result if sufficient suction is not maintained.

8. Magnetos — are checked by first noting the r.p.m. with the magneto switch in the BOTH position, the magneto switch is then moved *two* notches to the left (right magneto) and the r.p.m. noted. Next, the magneto switch is returned to BOTH and then switched *one* notch to the left (right magneto) and the r.p.m. noted. The magneto switch should then be placed back to the BOTH position for takeoff. This method reduces the possibility of the pilot inadvertently

taking off on one magneto. The permissible r.p.m. reductions, or "drops," are specified in the aircraft owner's manual.

The important thing to notice is the difference between the r.p.m. drops for the left magneto and the right magneto. One magneto should not show a drop of more than 75 r.p.m. compared with the drop of the other magneto. Different temperature and humidity conditions could cause a variation in r.p.m. drops for individual magneto operation from day to day. However, *each magneto* tested separately should indicate approximately the *same r.p.m. drop*. After the magneto check, the pilot should return the ignition switch to the BOTH position.

9. Carburetor heat — is pulled to the ON position and the power loss noted. With hot air entering the carburetor, the engine r.p.m. will drop, or decrease, indicating the carburetor heat *is functioning*.

10. Flight instruments and radios —

   a. The altimeter is set to the airport barometric reading as supplied by the tower controller, or set to indicate the correct field elevation if operating from an uncontrolled field.

   b. The heading indicator is set to coincide with the magnetic compass indication.

   c. All heading indicators are checked for stable operation.

   d. The OBS indicators are set to the desired course or radials.

11. Emergency Locator Transmitter — To determine if this unit has been activated accidentally, the pilot should tune his radio to 121.5. If the emergency locator transmitter is on, a loud buzy will be heard. If the unit is transmitting, return to the ramp and have the unit checked. If the unit is not transmitting, return the radio to its original setting.

When the checklist is complete, the pilot should call the tower for takeoff clearance.

If the pilot is operating from an uncontrolled field, a 360° taxiing turn in the direction of the traffic pattern should be performed to observe the entire traffic pattern for other traffic. Then, when the area is clear, the aircraft can be taxied to the takeoff position.

The Regulations state that aircraft on final approach to land, or landing, have the right-of-way over other aircraft in flight or operating on the surface. Good operating practice also requires that one aircraft should not take off over another aircraft that is completing its landing roll. At uncontrolled airports, the pilot should wait until the aircraft turns *off the runway* (or takes off again, if performing "touch-and-goes") before beginning his takeoff run.

TAXIING INTO TAKEOFF POSITION

When proper spacing has been determined, the pilot should taxi to the end of the runway, line up with the runway centerline, center the nosewheel, and neutralize the ailerons. After checking the windsock once again to determine the wind position in relation to the runway, the pilot is ready to begin the takeoff.

The right hand is placed on the throttle and should remain there throughout the takeoff. *Two exceptions to this* may be a rare occasion where the pilot needs two hands to move the control wheel or where a trim tab adjustment is needed. The pilot's hand on the throttle assures that it will not vibrate back during the takeoff roll and allows the pilot to quickly close the throttle should the decision be made to abort the takeoff roll.

The feet should be resting on the floor with the balls of the feet on the bottom edges of the rudder pedals. This will put the feet in a position where they will have no tendency to inadvertently press the toe brakes.

TAKEOFF PROCEDURES

The general procedures used for takeoff are as follows:

1. From a stationary position at the end of the runway, add power. Once the pilot begins applying power, the movement should be continuous up to the takeoff power setting, as though it were being "pushed through soft butter."

   As soon as power is applied, the aircraft will begin to roll. The pilot should pick out some object beyond the end of the runway in line with the centerline to use as a reference point for directional control.

2. Directional control is maintained with the *rudder pedals and not the control wheel*. The aileron control is used for wing position. If the wind is straight down the runway, the pilot should expect to hold neutral aileron. The rudder pedals, through nosewheel steering, are sufficient to maintain directional control even at slow speed; however, the rudder itself becomes more effective as speed increases. By the time the nosewheel is lifted from the runway, the airflow past the rudder is sufficient to provide directional control.

   The rudder pressures required during the takeoff roll can be anticipated as shown in figure 3-2. At the starting position, the rudders are essentially neutral. As power is added, right rudder pressure is applied to counteract the

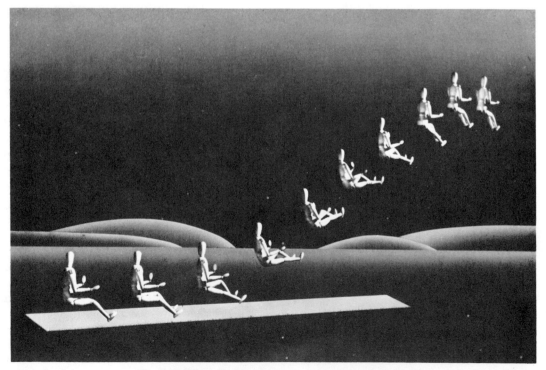

Fig. 3-2. Right Rudder Pressure Varies During Takeoff

engine torque effect. When the airplane is placed in the climb attitude, the right rudder pressure must be increased to compensate for propeller P-factor caused by high angles of attack. The right rudder pressure (which is exaggerated in the illustration) is required throughout the climb until the aircraft has reached level flight cruise conditions.

3. As speed increases, apply slight back pressure to the control wheel. The airplane will continue to roll with the nosewheel on the ground. With the increase in speed, the controls will become increasingly more responsive. Rudder movement, combined with nosewheel steering, will easily control the nose movement to the left or right.

4. As speed continues to increase, elevator responsiveness will become sufficient to make the nosewheel light on the ground. This will occur before the aircraft has sufficient speed to become airborne. At that time, the *takeoff attitude is established.* This attitude is slightly nose-high and the nosewheel may be just barely off the ground. In most training aircraft, this attitude is similar to the *normal climb attitude.*

This attitude is important because it is a compromise between holding the nose on the ground and selecting an attitude which is too nose-high. With the nose on the ground, the airplane tends to build excess airspeed, which increases the length of runway required for takeoff. With an excessively nose-high attitude, the aircraft may

a. be forced into the air prematurely and, therefore, settle back to the runway again, or

b. be at such a high angle of attack (high drag condition) that the aircraft will not be able to rapidly accelerate to the climb airspeed.

With the proper attitude, the aircraft will attain a safe speed and become airborne when it is ready to fly. Simply holding that attitude after becoming airborne will allow the aircraft to readily accelerate to climb speed. Another advantage of maintaining this attitude is that if the aircraft is prematurely lifted into the air by a gusty wind, it will be in the best attitude to accelerate to climb speed, or if the aircraft settles, it can safely touch the runway and continue the takeoff roll.

5. When the aircraft reaches climb speed, adjust the nose attitude to the proper climb attitude (if not already at that attitude). When all obstacles have been cleared, reduce the power to the normal climb setting. The aircraft should now be established on the takeoff leg of the traffic pattern and flown straight out on an extension of the centerline of the runway without drifting to one side or the other.

6. When the airplane has reached at least 400 feet AGL (this altitude depends on atmospheric conditions, traffic, and the airport layout), the turn to crosswind may begin unless other traffic requires additional spacing. Look to the left and behind to ascertain that the area is clear of traffic before commencing the turn to the crosswind leg. When the aircraft is established on the crosswind leg, continue the climb and depart the pattern in the prescribed manner.

CROSSWIND TAKEOFF

One of the factors to be considered in the practice of takeoffs is the effect of a crosswind. The instructor will explain the maximum safe crosswind velocities (crosswind component) given in the owner's manual (if listed) for that particular aircraft. The term "crosswind component" refers to *that part of the wind velocity that acts at right angles to the aircraft's path on takeoff or landing.*

The chart excerpted from a typical owner's manual, shown in figure 3-3, indicates a maximum safe direct crosswind of 17½ m.p.h. It is possible to take off with the wind velocity higher than 17½ m.p.h. providing the crosswind component does not exceed 17½ m.p.h. For example, a 25 m.p.h. wind acting at an angle of 30° to the runway does not place the aircraft in danger of being upset or forced off the runway. This is illustrated by the intersection of the vertical and horizontal dashed lines in figure 3-3. The chart clearly shows that the smaller the angle of the wind to the runway, the smaller the danger from an excessive crosswind component.

During ground operations such as the takeoff and landing roll, a crosswind can be thought of as attempting to push the aircraft to the downwind side of the runway, or to get under the wingtip and roll the aircraft to the downwind side, as explained in the discussion of taxiing in Chapter 1. To compensate for this effect, the aileron controls are used as if to turn or roll the aircraft into the wind as in taxiing operations.

When the aircraft is rolling slowly along the runway, the effect of the aileron is slight, so *full aileron deflection* should be used. The wind flow over the wing with the up aileron tends to hold that wing down. Airflow under the wing with the down aileron tends to push that wing

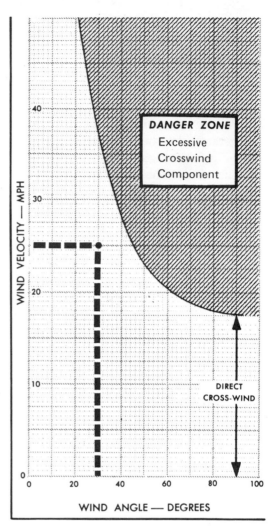

Fig. 3-3. Maximum Safe Crosswind Velocity Chart

up, thus counteracting the tipping tendency. (See Fig. 3-4.)

As flying speed is approached, the ailerons become more effective and the aircraft will bank into the wind and roll along on one wheel. (See Fig. 3-5.) Since an airplane in a left bank also tends to turn left, the pilot will find it necessary to apply enough right rudder pressure to keep the nose of the aircraft pointing straight down the runway.

There is another reason rudder pressure is needed on the downwind side of the

Fig. 3-4. Aileron Positions During Crosswind Takeoff Roll

aircraft. In ground operations, the crosswind is attempting to "weathervane" the aircraft. The weathervaning tendency, illustrated in figure 3-6, is the result of the large rudder surface sticking up into the airflow at some distance from the center of gravity which tends to move the nose into the wind.

Fig. 3-5. Rolling On One Wheel During Crosswind Takeoff

Fig. 3-6. Weathervane Tendency

Determining how much aileron and rudder is required to compensate for a crosswind requires a little trial and error practice. When the right amount is obtained, the pilot will feel *no side load and the aircraft will track straight down the runway.*

The general technique to use in a crosswind takeoff is the same as the normal takeoff except for consideration of the side wind.

1. Line up the aircraft with the centerline of the runway and apply full aileron into the wind, as shown in figure 3-7.

2. Apply power in the normal manner. Then, as speed increases and ailerons become effective, reduce aileron pressure until there is just enough to counteract the crosswind.

3. As flying speed is approached, raise the nose to takeoff attitude

A crosswind takeoff in light winds requires small corrections. In moderate winds, the aircraft will feel sensitive to side loads when it is at the transitional stage where it has enough speed and lift to be light on the wheels, but not enough to be airborne. If the corrections are improper, side loads may be placed on the landing gear.

Fig. 3-7. Beginning The Crosswind Takeoff Roll

causing the aircraft to roll up on one wheel. Use rudder as necessary to prevent weathervaning and to keep the nose straight down the runway. (See Fig. 3-8.)

When the aircraft is airborne and clear of the runway, the wings are rolled level and the aircraft pointed into the wind, or *crabbed*, so that the aircraft tracks straight out on an imaginary extension of the runway centerline. (See Fig. 3-9.)

An alternate method of accomplishing a takeoff in a crosswind or gusty wind condition is to hold the aircraft on the runway until a slightly higher than normal lift-off speed is attained. At this point, the aircraft is lifted abruptly off the runway and established in a normal climb attitude. This technique reduces the chance of the aircraft being prematurely lifted off the runway by a sudden gust of wind before the aircraft has attained sufficient airspeed to remain airborne. Care should be taken *not* to lift the main landing gear off the runway leaving only the nose gear touching. This is called "wheelbarrowing" and is the result of excessive airspeed and forward control pressure.

"ACCEPTABLE TAKEOFF PERFORMANCE"

Acceptable performance for takeoff is shown by proper use of checklists, prompt and smooth application of power, positive directional control, and recognition and use of the takeoff attitude. Wind drift correction, directional control, and the application of proper techniques must be positive and correct. The student should not pull the aircraft off the ground too soon or remain on the ground too long.

SIMULATED TAKEOFF EMERGENCIES

With today's reliability in engine operation, an engine malfunction on takeoff is

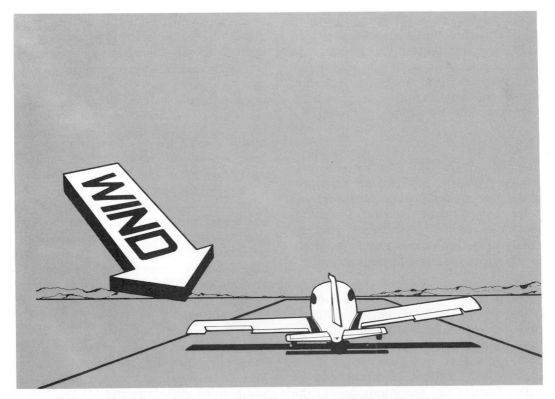

Fig. 3-8. Aircraft Attitude At Liftoff During Crosswind Takeoff

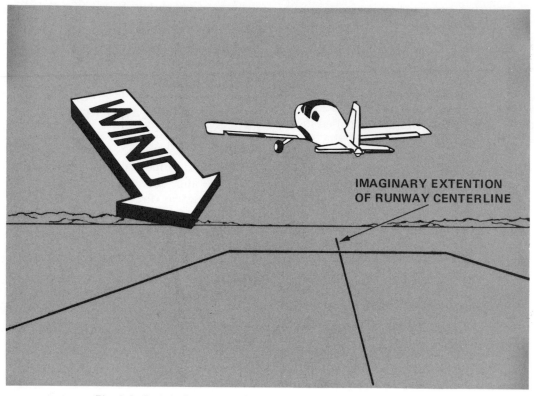

Fig. 3-9. Crab Is Required After Lift-off To Counteract Crosswind

rare; however, since the remote possibility of a malfunction does exist, the flight instructor will present this situation to the student by simulating an engine failure. The simulated emergency situation will occur when the instructor reduces engine power by suddenly closing the throttle.

The objective of this practice is to prepare the student pilot to meet the problems of a forced landing on takeoff. It helps the student develop judgment, technique, and confidence, and gain the ability to respond quickly and accurately to situations requiring prompt action.

The first factor that must be considered is the possibility of an engine malfunction on the takeoff roll or just after "breaking ground." It was mentioned previously that the pilot's right hand should be on the throttle throughout the entire takeoff roll and climbout. On the takeoff roll, if the pilot suspects a mal-

function, his hand is in position to quickly abort the takeoff by moving the throttle to the idle position. The brakes are used as necessary to slow down and the rudder is used to maintain directional control.

If the aircraft has become airborne to a height of approximately 10 feet and the instructor initiates a "simulated" emergency by "pulling off" the power, the student should hold the present attitude and allow the aircraft to settle back on the runway. If the loss of power is simulated after climbing above 10 feet, the nose should be lowered to the approach attitude and a flareout begun at normal flare altitude. The rule of thumb is that below 400 feet AGL, if power is lost, the pilot has no choice but to continue straight ahead or make only very small changes in direction to avoid obstacles. This is because not enough altitude is available to make a power-off 180° turn back to the airport and land.

At the completion of the maneuver, power should be added immediately to initiate a go-around. If the instructor says, "I've got it," the student should remove his hands and feet from the controls immediately.

Under the instructor's guidance, the new student will be exposed to and trained in as many possible situations as the instructor can anticipate. Many or all of the simulated emergencies practiced will never occur, but the practice received will be valuable in learning the characteristics of the aircraft and building confidence so that when the student is alone, he is able to handle the aircraft within a wide variety of situations.

SECTION B—NORMAL AND CROSSWIND LANDINGS

APPROACH AND LANDING

The approach and landing is the phase of flight most student pilots look forward to because it represents a high degree of accomplishment. Success in landings places the student one step closer to the exciting day when he can solo the aircraft and begin to think of himself as a pilot. Success in landings *does not eliminate* the need for further flight instruction, but it is an *important step* because it is an indication of the student's ability to execute basic maneuvers, plan and think ahead of the aircraft, and divide his attention among several tasks.

The landing itself is nothing more than the transition of the aircraft from an airborne machine to a ground machine. The objective of a landing is to make a safe transition between the air and the ground. To be successful, one must use procedures and techniques that consistently land the aircraft on the desired area of the landing surface without violating good safety practices.

The successful landing is the result of planning that begins on the downwind leg. For example, consider the problem of touching down in a preselected area, or point, on the runway and all the things that vary or influence the aircraft during the approach to that point. (See Fig. 3-10.)

The distance traveled from "A" to "C" will vary with the aircraft's distance from the runway on the downwind and base legs. Wind and airspeed influence how fast the aircraft will travel that dis-

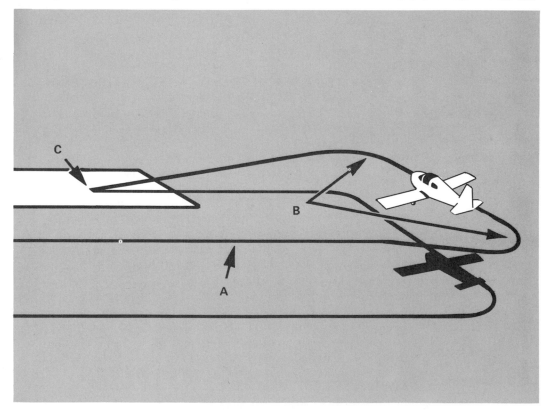

Fig. 3-10. Spot Landing

tance while power, airspeed, and flaps influence the rate at which the aircraft descends.

If, on each approach, the pilot uses a different distance from the runway, airspeed, flap setting, or a different rate of turn to base and final, he has a new problem to solve on each approach and his chances of consistently touching down at the appropriate spot on the runway are considerably reduced. Thus, one goal in the traffic pattern and in approach-to-landing practice is to eliminate as many of the variables as possible.

An attempt should be made to funnel the aircraft to one place. (See Fig. 3-11.) For example, at point "A" the distance of the downwind leg from the runway should be consistent. Also, the altitude, the point at which power is reduced, and the approach airspeed should be the same for each approach.

The turn to base and final at the points labeled "B" in figure 3-12 should be consistent; then, flaps and power can be used to make minor corrections for position and wind influences. Throughout the instructional period, the instructor will be urging the student to be consistent and exercise precise control of the aircraft to reduce the number of variables in the landing approach.

The techniques and procedures used in the approach and landing will be somewhat different in various makes and models of aircraft. The instructor will refer to the owner's manual and then, with the aid of his own experience, demonstrate how the approach and landing will be performed in that aircraft. A typical approach and landing is described in the following paragraphs.

DOWNWIND LEG

The downwind leg should be flown at a distance of approximately 2,500 feet, or one-half mile, from the runway in use. The aircraft's track should parallel the runway with no tendency to angle toward or drift away from the runway. (See Figure 3-13.) Angling or drifting causes the traffic pattern to have an abnormal shape which greatly influences the length of the base leg.

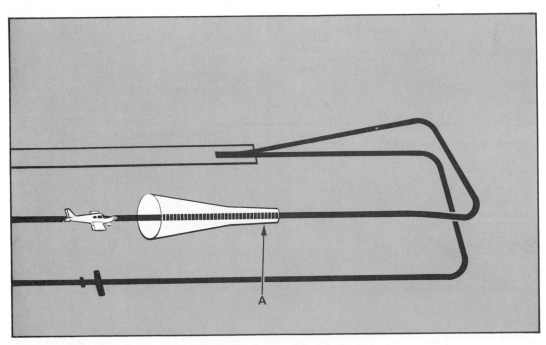

Fig. 3-11. Funnel Aircraft To One Spot

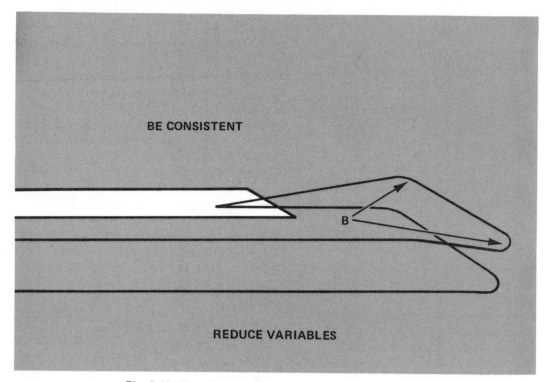

Fig. 3-12. Turn To Base And Final Should Be Consistent

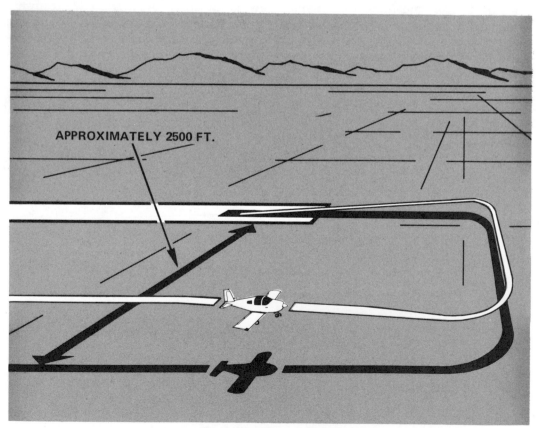

Fig. 3-13. Downwind Leg

Airspeed is usually near cruise in light trainers (or at a reduced cruise speed for higher performance aircraft) when approaching the point opposite the intended landing, *or the 180° point* as it will be referred to from now on. Altitude should be at the designated traffic pattern altitude. (See Fig. 3-14.)

180° POINT

Just before reaching the 180° point, the pilot should perform the prelanding checklist. Then, at the 180° point, the power is reduced to the usual descent power setting (this could be idle r.p.m. or a very low power setting), altitude maintained, and the airspeed allowed to slow to approach speed. (See Fig. 3-15.)

WING FLAPS

When flying slower than the maximum flap extension speed (shown by the high-speed end of the white arc on the airspeed indicator), the flaps may be lowered. Initially, the instructor will suggest how much flap extension to use and when to lower the flaps, although a one-half setting is most often used in the early stages of training. (See Fig. 3-16.)

BEGINNING THE DESCENT

When the aircraft has reached the approach airspeed, that speed should be maintained and descent initiated. The transition from cruise speed on downwind to the descent speed is a practical application of the transition from cruise to descent that has been introduced in the practice area. The recommended descent or approach speeds are found in the owner's manual for the aircraft.

If an approach speed range is given, for instance from 70 to 80 m.p.h., it is recommended that the pilot select one speed, such as 75 m.p.h., as the "target" and maintain an attitude that will produce that speed as precisely as possible. An acceptable rule of thumb is that the final approach speed is 1.3 times the power-off stall speed in landing configuration. In any case, precise control over aircraft attitude will result in control of the airspeed.

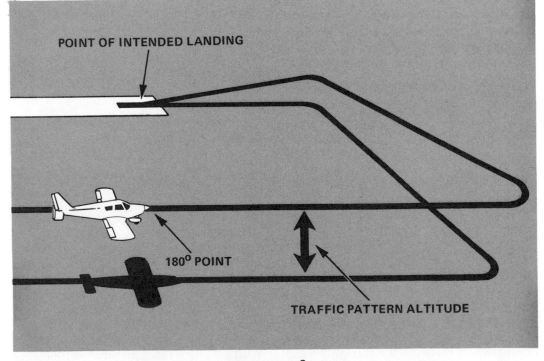

POINT OF INTENDED LANDING

180° POINT

TRAFFIC PATTERN ALTITUDE

Fig. 3-14. 180° Point

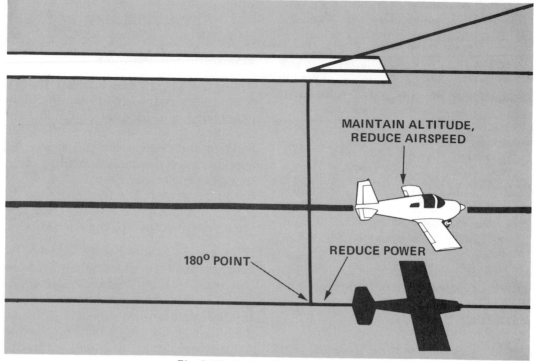

MAINTAIN ALTITUDE,
REDUCE AIRSPEED

REDUCE POWER

180° POINT

Fig. 3-15. Reduce Power And Airspeed

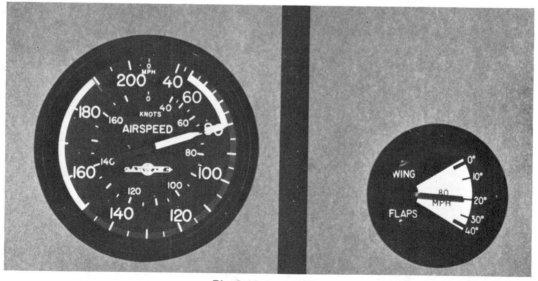

Fig. 3-16. Lower Flaps

BASE LEG

The turn to base leg usually begins after the aircraft has descended approximately 100 to 200 feet; however, the principle cue of when to begin the turn will come from sighting the aircraft's position re-lative to the runway. The runway should appear to be between 30° and 45° be-hind the wing and look similar to the view shown in figure 3-17.

With practice, this "picture" will become so familiar that the student can detect

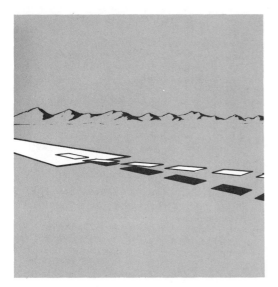

Fig. 3-17. Begin Turn To Base Leg

the need for small corrections at this position. The start of the turn should be varied to compensate for variations in conditions. For example, if the downwind leg is wider than normal, the turn should be started a little sooner, as shown in figure 3-18. As the new student

gains experience, he will be better able to judge these variations. He also should be alert for other traffic in the pattern. The pilot always should check for aircraft on base leg below him, above him, or on an extra long final approach.

KEY POSITION

When the aircraft rolls out on base leg, it has arrived at a major decision point called the *key position*. (See Fig. 3-19.) On each approach to a landing, when the key position is reached, the pilot must assess his position and determine whether or not corrections must be made to the approach pattern. This assessment will be based on four factors: altitude, airspeed, distance from the runway, and wind.

For example, if the aircraft is close in, as shown in position 1, in the top portion of figure 3-19, or high (position 2), the pilot will land beyond the desired touchdown point.

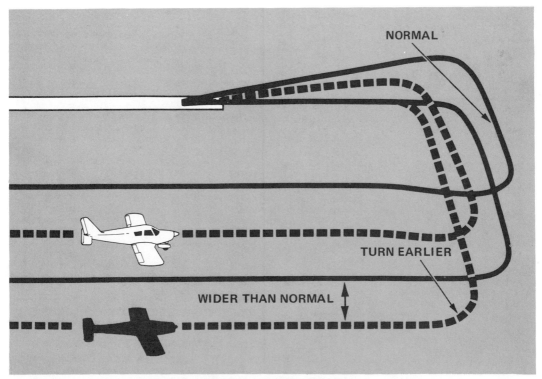

NORMAL

TURN EARLIER

WIDER THAN NORMAL

Fig. 3-18. Wider Than Normal Pattern

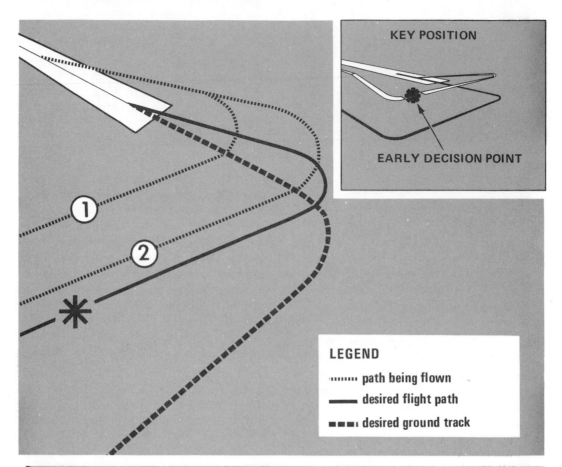

KEY POSITION

EARLY DECISION POINT

① ②

LEGEND

∙∙∙∙∙∙∙∙ path being flown

——— desired flight path

▬ ▬ ▬ desired ground track

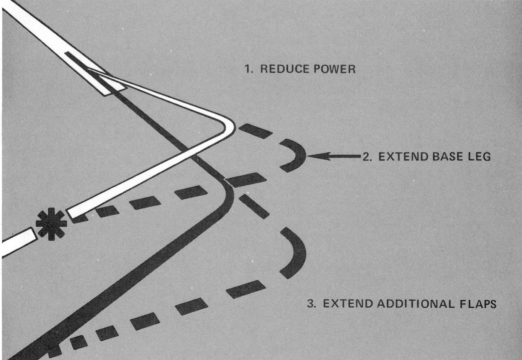

1. REDUCE POWER

2. EXTEND BASE LEG

3. EXTEND ADDITIONAL FLAPS

Fig. 3-19. Key Position Is Decision Point

TAKEOFFS, LANDINGS, AND SLIPS

There are three possible corrections the pilot can make: (1) reduce power *further*; (2) extend the base leg *farther* out; or (3) extend *additional* flaps. (See Fig. 3-19, bottom portion.) Any one or all three can be used, depending on how large a correction is necessary.

If the aircraft is low (see figure 3-20, position 1), or wide on base leg (position 2), or the wind stronger than normal, the pilot will land short of the desired point if no corrections are made.

In these cases, there are two actions he can take: (1) begin the turn to final *sooner* than normal, or (2) *add power*. Either one, or both, may be used, depending on the amount of correction necessary. Retracting the flaps is usually not considered an acceptable correction. Normally, once flaps are extended, they are not retracted until the landing has been completed, or the landing approach abandoned.

The corrections for position should be performed any time the pilot recognizes a need for a correction. The significance of the key position is that it is an early decision point where the pilot can easily make major adjustments to insure a smooth approach and avoid large or violent last-minute corrections. Throughout the approach, the pilot should continue to assess his position relative to the runway to determine the need for corrections. Through the process of assessing his position, making corrections, and then reevaluating the touchdown point, the pilot will be able to judge his touchdown point accurately.

FINAL APPROACH

Before turning to the final approach leg, the pilot *always* should look in all directions for other traffic, since an aircraft on final has the right-of-way over an aircraft on base leg. If the area is clear, the turn to final then can be made. The turn to final should be a coordinated

turn and is usually completed between 200 and 400 feet AGL.

The turn to "final" also should be planned so the airplane rolls out on an extension of the runway centerline. The final approach should neither angle to the runway, as shown by the aircraft on the left in figure 3-21, nor require an "S-turn," as shown by the aircraft on the right. Approach airspeed should be maintained in the turn, on final approach, and monitored until beginning the landing transition.

Throughout the landing approach, the student's ability to use the *visual attitude references* to control speed is the key to a good landing. The landing approach is a combination of dividing attention between looking outside the cockpit at the runway and other aircraft traffic and accurately controlling the speed. The student is encouraged to look outside since, if he can judge his *airspeed through his aircraft attitude*, he will exercise better speed control and perform smoother approaches. The nose attitudes should look familiar since they are the ones used when practicing straight descents and descending turns. As a reminder — *elevators primarily control the nose attitude which in turn controls airspeed.*

It is quite common during early landing practice for the student to overcontrol and make corrections that are in excess of the required amounts. As a guide, the student should visualize the normal or standard pattern and make positive smooth corrections back to that pattern. If he is low, for example, it usually is not necessary to make an abrupt correction and climb back to the standard pattern, as shown by path number 1 in figure 3-22. Rather, he can reduce the rate of descent or go to level flight for a short period, as shown by path number 2.

If the aircraft is "low and slow," power must be added and the nose lowered.

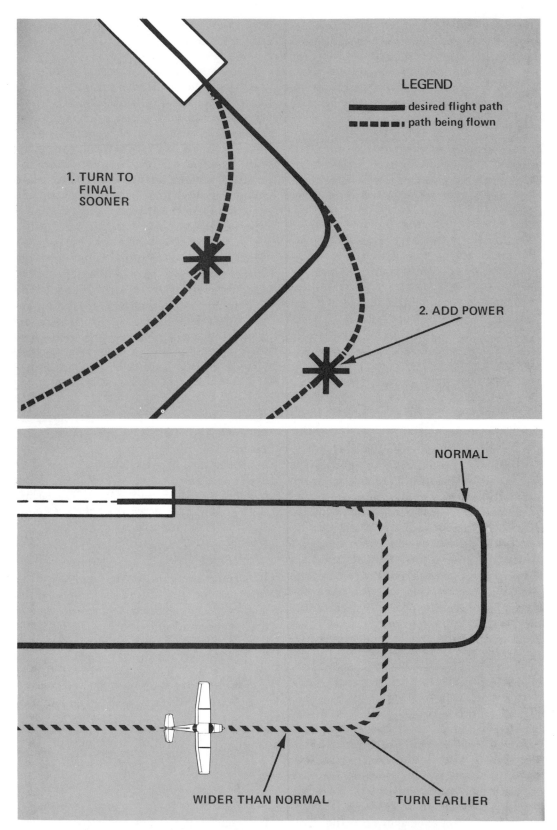

LEGEND

━━━ desired flight path
■ ■ ■ ■ path being flown

1. TURN TO
   FINAL
   SOONER

2. ADD POWER

NORMAL

WIDER THAN NORMAL

TURN EARLIER

Fig. 3-20. Corrections When Flying Too Low Or Too Wide On Base Leg

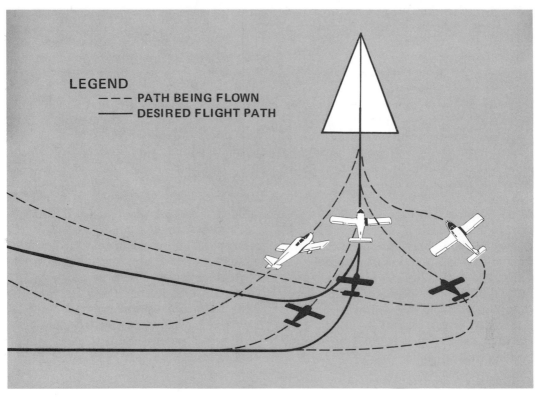

Fig. 3-21. Turn To Final Should Produce Roll Out On Runway Centerline

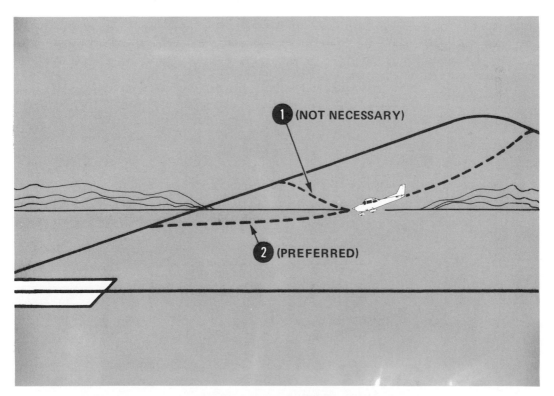

Fig. 3-22. Low Approach Correction

Lowering the nose when the aircraft is low is sometimes a difficult response that must be practiced and learned.

Approach speed is frequently near the best angle-of-glide speed. This means that any airspeed, either lower or higher, will result in a higher rate of descent. If the aircraft is short of the desired touchdown spot, the pilot *cannot* stretch a glide by raising the nose and slowing the airspeed; instead, *power must be added.*

If the approach is extremely high, there is very little value in diving the aircraft. It may look, at first, like the aircraft will "make" the desired point of touchdown, but in the dive, excessive airspeed is built up which *must be dissipated* in the flareout. The result is that a greater distance is covered as the aircraft floats down the runway, usually well beyond the desired landing point.

LANDING

The landing consists of a flareout (to reduce speed and decrease the rate of descent), the touchdown, and the rollout. (See Fig. 3-23.) The term "flareout" is used throughout this publication to mean the process of changing the attitude of the aircraft from a glide or descent attitude to a landing attitude. Other terms used interchangeably for this maneuver are: *flare*, *roundout*, and *level-off*.

The point at which the pilot will actually touch down can be estimated by finding the point where the aircraft glide path intersects the ground and adding the approximate distance to be traveled in the flareout. The glide path intersection is the point on the ground that has *no apparent relative movement*. As the aircraft descends, all objects beyond the glide path intersection point will appear to move *away* from the aircraft, while objects closer will appear to move *toward*, or under, the aircraft. Only the glide path intersection point will appear to remain in the same relative position. With some experience, this point will be easy to locate.

From the glide path intersection point, the approximate distance traveled in the flareout will vary somewhat depending upon the make and model of the aircraft, approach speed, and wind. For example, if the pilot wishes to touch down at the point shown in figure 3-23, his apparent aiming point for the glide path

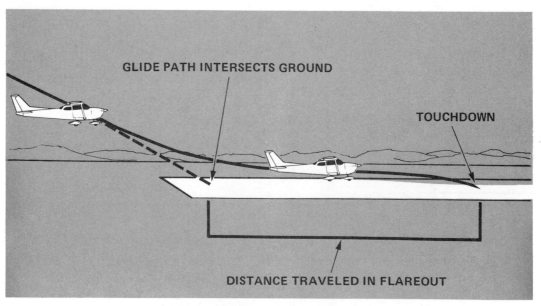

GLIDE PATH INTERSECTS GROUND

TOUCHDOWN

DISTANCE TRAVELED IN FLAREOUT

Fig. 3-23. Flareout Distance

should be approximately 150 feet short of that point.

The flareout begins at different altitudes for aircraft of varying weights and approach speeds; however, for most training aircraft, it begins at approximately 15 feet AGL. The flareout is initiated with a gradual increase in back pressure on the control wheel to reduce speed and decrease the rate of descent. Ideally, the aircraft will reach zero rate of descent approximately one foot above the runway at about 10 miles per hour above stall with idle power. (See Fig. 3-24.)

Then, the pilot simply attempts to *hold the aircraft at the one-foot altitude* by continuing to increase the back pressure. The aircraft will slowly settle to the ground in a slightly nose-high attitude as it approaches stall speed. The nose atti-

tude at touchdown should be very close to the nose attitude at takeoff, with no weight on the nose gear at touchdown. *Back pressure* on the control wheel *should be maintained until the nose-wheel touches down*, and then slowly relaxed. (See Fig. 3-25.)

The clues that the pilot uses in the flareout and landing are a combination of visual and kinesthetic feelings. Descents and approaches to stalls have been practiced to build sensitivity to control responses and smoothness in preparation for the flare and landing. Generally, kinesthetic (or "seat of the pants") sensitivity is not fully developed at the time landing practice begins; thus, vision is the most important sense.

The altitude to begin the flareout and the height throughout the flareout is de-

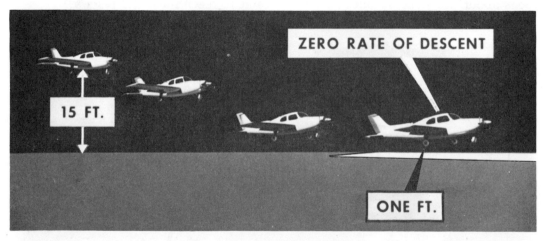

Fig. 3-24. Initiation of Flareout

Fig. 3-25. Attitude At Flareout

termined by the pilot's depth perception. Depth perception is the visual comparison of the size of known objects on the ground, so it is important *where* the pilot focuses his vision during the approach. If he focuses *too close* to the aircraft, or looks straight down, the airspeed will blur objects on the ground causing his actions to be too abrupt or delayed. The tendency will be to overcontrol, level-off high, and make a "pancake landing."

If the pilot focuses *too far* down the runway, he will be unable to accurately judge his height above the ground and consequently his reactions will be slow, since there will seem to be no cause for action. In this case, the pilot will "fly into the ground." Thus, it becomes obvious that the pilot must focus at some intermediate point. A guideline is to focus about the same distance ahead of

the aircraft as one would in a car going the same speed. (See Fig. 3-26.)

During the transition of the aircraft from an airborne machine to a ground machine, the aircraft should be pointed straight down the runway. When flying a side-by-side aircraft, the reference point for sighting over the nose *is not* the center of the nose, but a point *directly in front* of the pilot's eyes. If the pilot sights over the center of the nose, he will tend to line up the aircraft in a crab. The touchdown will then place heavy side loads on the landing gear.

Directional control is maintained during the rollout with rudder pressure. On landing, the feet should be in the same position on the rudder pedals as they were in the takeoff. With the heels on the floor, there is no tendency to inadvertently use the brakes, but the pilot is able to reach them if necessary.

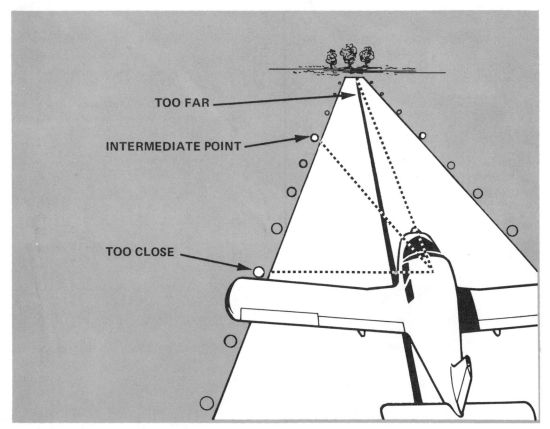

Fig. 3-26. Focus Point For Flareout

GO - AROUND

A general rule of thumb used in landing practice is that if the aircraft has not touched down in the *first third of the runway*, the pilot should abandon the landing by flying around the traffic pattern and setting up another landing approach. A go-around may also be required when another aircraft, truck, or some other obstacle is on the runway, or whenever the pilot feels that the approach is uncomfortable, incorrect, or potentially dangerous.

The decision to make a go-around should be positive and made before a critical situation develops. When the decision is made, it should be carried out without hesitation. In many cases, the go-around is started from a slow airspeed and nose-high attitude. The first response should be to apply all available power (includes turning off carburetor heat, if used) and adjust the pitch attitude to a normal climb attitude; however, the aircraft should be accelerated to the best angle-of-climb speed *before* the climb is started, and the flaps should be retracted slowly. Flap retraction always should be done with care since, at low speeds, it is possible for the aircraft to stall or lose altitude if the flaps are retracted too rapidly.

To initiate a go-around after the aircraft has reached a safe speed, a gentle turn should be made to the right side of the runway. (See Fig. 3-27, top portion.) Special cases, such as parallel runways, etc., may prevent this slight turn or may require a slight left turn. The flight path should be far enough to the side of the runway so as not to interfere with an aircraft taking off. In this position, the pilot can clearly see the runway and thus avoid flying directly over any traffic taking off.

The climb on the takeoff leg should be continued parallel to the runway until the crosswind leg is reached. If other aircraft are ahead, the pilot should allow

proper spacing and rejoin the traffic pattern. Care also should be exercised to avoid aircraft that may be taking off and climbing out close behind. (See Fig. 3-27, bottom portion.)

BOUNCED LANDINGS

In landing practice, the student may experience some bad landings and bounce into the air. Usually, it is wise not to attempt to "salvage" the landing; rather, the pilot should make an immediate go-around. Except when lowering flaps or trimming the aircraft, the student should have his hand on the throttle at all times throughout the traffic pattern approach and landing. When a landing is bad, a decision should be made to go around and all available power added. The nose then is adjusted to a normal climb attitude and the go-around procedures carried out. The only exception to the previously described procedure is that the student can climb straight ahead.

If, on landing, there is just a little skip and only a few feet of altitude gained, the nose should be placed in the landing attitude and held in that attitude while the aircraft settles back to the runway. The same procedure should be used if the aircraft "balloons" because the student has pulled back on the elevator control too rapidly. The nose should be replaced in the landing attitude and held in that attitude until the aircraft settles on the runway. If there is any question about the successful outcome of the landing, the best procedure is to go around.

USE OF WING FLAPS DURING LANDING

In addition to the use of partial wing flaps, full-flap and no-flap landings, together with a few variations, also will be demonstrated and practiced by the student pilot. The procedures used with various flap settings are generally the same; however, there are specific differences of which the pilot should be aware.

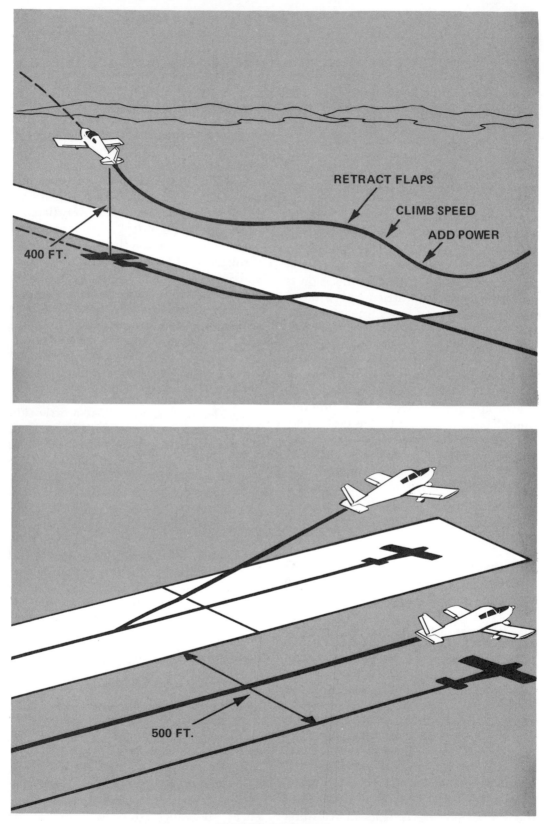

Fig. 3-27. Go-Around

With *no flaps*, the rate of descent usually is less than with half flaps, so the student tends to be a little high and *land long*. To avoid this, power and the traffic pattern must be adjusted. Since the no-flaps stall speed is higher than with full flaps, the aircraft touches down at a faster speed and the runway rollout is longer.

For full-flap landings, the same procedures are utilized as in the half-flap approach. Three-fourths flaps are usually extended on base leg. Normally, the extension to the full-flap position is performed on the final approach leg. (See Fig. 3-28.)

At the *full-flap setting*, the rate of descent increases so the tendency is to *land short*. The touchdown speed is slower since the full-flap stall speed is less than the partial - or no-flap stall speed. This also results in the ground roll being shorter.

If a go-around is required with full flaps, the pilot should follow the procedures previously described. After adding all available power and removing carburetor heat, the pilot should not immediately attempt to go to a normal climb attitude. Instead, he should go to the straight-and-level, full-flap, slow flight attitude as this will allow the aircraft to maintain its present altitude. As the best angle-of-climb speed is approached, the flaps may be raised slowly. As the aircraft accelerates and the airspeed gradually increases, the nose attitude should be adjusted to the climb attitude. An attempt to immediately raise the nose in a full-flap go-around may result in a stall; or suddenly raising flaps may cause the aircraft to descend onto the ground.

CROSSWIND LANDING TECHNIQUE

The approach to landing in a crosswind is essentially the crosswind takeoff process in reverse. The turn to final should be completed on an extension of the runway centerline with the aircraft in a crab to correct for wind drift. (See Fig. 3-29.)

On final, in preparation for the landing, the wing is lowered into the wind. To prevent the aircraft from turning, *opposite rudder pressure* is used to keep the nose pointed straight down the runway. This condition is similar to a *side slip* which will be discussed in a later section. When properly performed, there is no tendency to drift from one side of the runway to the other. Beginning pilots usually place the aircraft in the wing-low condition at altitudes of 100 to 200 feet while on final, but as they become more experienced, they tend to wait until just before beginning the flareout. (See Fig. 3-30.)

At the normal altitude for flareout, the flare is started. The wing is held down throughout the flare and touchdown, causing the aircraft to touch down on one wheel, as shown in figure 3-31. The airplane must contact the runway *without drifting to either side*. *Both* the ground track and the longitudinal axis of the airplane must be aligned with the runway when the airplane contacts the ground; otherwise, extreme side loads will be imposed on the landing gear and tires resulting in damage. A good rule of thumb to remember is: *control the drift with aileron and the heading with rudder.*

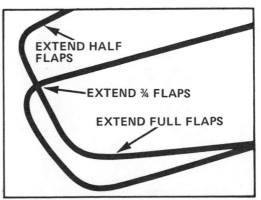

Fig. 3-28. Full-Flaps Extension Sequence

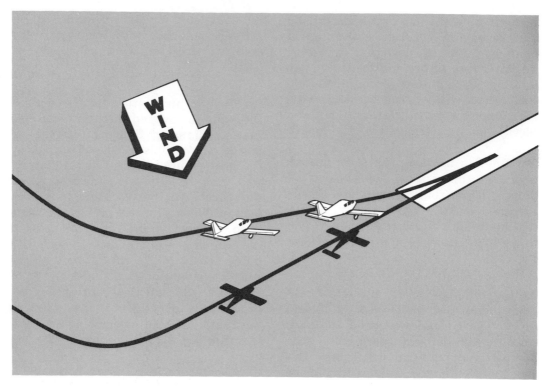

Fig. 3-29. Crosswind Final Approach

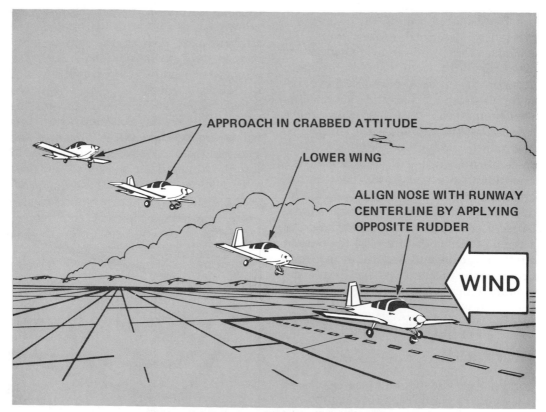

Fig. 3-30. Crosswind Landing Preparation

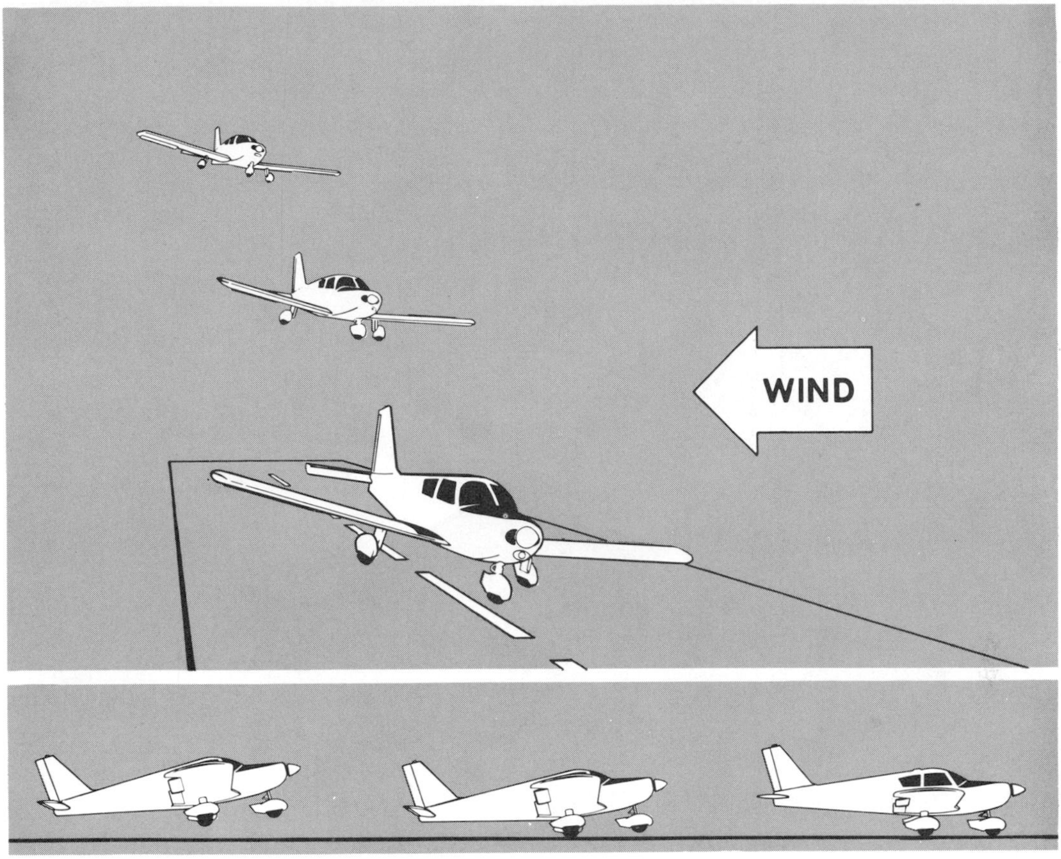

Fig. 3-31. Crosswind Landing Flareout

As the airplane slows, the downwind wing will lower and the other wheel will touch the ground. As the aircraft continues to slow, more aileron is added, since the aileron becomes less effective with speed loss.

In moderate and light crosswinds, the nose attitude for lift-off and touchdown is the same as in normal takeoffs and landings. In a strong crosswind, the nose is placed on the ground with positive elevator pressure to allow the nosewheel to assist in directional control. (See Fig. 3-32.) Care must be taken not to lift the main landing gear off the runway as a result of excessive airspeed and forward control pressure. When the speed has slowed and the nosewheel touches the ground, positive forward elevator pressure assists in maintaining directional control; in addition, aileron controls always should be *fully* into the wind (not

neutralized) at the end of the landing roll.

In gusty wind conditions, a slightly different landing technique may be used. Better control can be maintained if, instead of landing the aircraft in a stalled condition, the aircraft is flown onto the runway at an airspeed slightly higher than normal and then held on the runway with *slight* forward elevator pressure.

In a left-hand traffic pattern, when the wind is from the right side of the runway, the wind tends to distort the pattern by pushing the aircraft wide on the downwind leg and slowing the groundspeed on the base leg. This condition, if not compensated for, will result in approaches that are *short* of the runway and shallow, angling turns to final.

Fig. 3-32. Control Positions For Crosswind Landing Rollout

When the wind is from the left of the landing runway heading, it tends to push the aircraft closer to the runway and to increase groundspeed on base leg. This situation tends to cause *overshooting* on final and approaches that are *high and long.* Recognizing these tendencies, the pilot should make compensations in the traffic pattern.

ACCEPTABLE PERFORMANCE FOR LANDING

Acceptable performance in the landing demonstrates recognition of the normal traffic pattern and correct attitude by using outside visual references to main-tain the approach speeds. Corrections for varying conditions on the approach and for bad landings should be prompt, posi-tive, and exact. Touchdown should be reasonably close to the desired point and on the runway centerline. The aircraft's longitudinal axis should be pointed straight down the runway and the air-plane should not drift due to a cross-wind. Correct crosswind techniques should be demonstrated throughout the touchdown and landing rollout. The student should display proper wind drift correction, power technique, directional control, judgment, and smoothness throughout the approach and roll-out.

SECTION C – SLIPS

The slip is a flight attitude used to increase the angle of descent without causing the airspeed to increase. The basic purpose of the slip is to expose as much of the airplane surface to the onrushing air as possible, so that the airplane frontal area produces considerable drag. (See Fig. 3-33.) This allows a

Fig. 3-33. Slip Attitude

steeper angle of descent without any acceleration. Flaps serve the same purpose but they cannot always be used because of crosswinds or gusts, so a knowledge of slips is useful.

There are two types of slips — the *forward slip* and the *side slip*, but they are performed using similar procedures. The forward slip is a maneuver used to cause high drag by flying partly "sideways" which results in an increased rate of descent. This maneuver is often used during an approach when the pilot is lined up with the runway but is *above* the normal glide path.

The side slip is a maneuver normally used when landing in strong crosswinds, since the longitudinal axis remains parallel with the direction of flight. The two types of slips are aerodynamically the same, but they differ in the way in which the airplane is maneuvered with respect to the ground.

FORWARD SLIP

To initiate a forward slip, one wing is lowered slightly and opposite rudder is applied simultaneously to keep the airplane on the same ground track. (See Fig. 3-34.) This procedure will keep the aircraft on a course which is aligned with the extended centerline of the runway, but the longitudinal axis of the airplane will be angled away from the runway. As soon as sufficient altitude is lost, all the controls should be coordinated again for a normal glide, flareout, and landing.

The forward slip can be valuable when landing in fields where obstructions may be encountered since the pilot has an excellent view of the landing area during the entire slip. In an aircraft with side-by-side seating, it usually is more comfortable to slip toward the side on which the pilot is sitting since the structure of the airplane provides something to lean against. Also, it is likely that the range of vision will be much better if the slip is made to that side. In airplanes with a tandem seating arrangement, a slip to

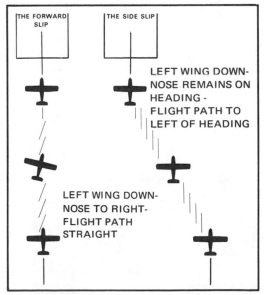

Fig. 3-34. Forward And Side Slip --
No Crosswind Present

either side may be used. If there is any crosswind, the slip will be much more effective if made toward the wind.

The only difference between the control operations in right and left slips is that the control pressures are reversed. Forward slips should be done with the engine throttled since there is little logic in slipping to lose altitude if the power is on.

To place the aircraft in a forward slip, the wing on the side toward which the slip is to be made is lowered by use of the ailerons. (See Fig. 3-35.) Simultaneously, the nose is swung in the *opposite* direction by use of opposite rudder so that the longitudinal axis is at an angle to the flight path. The nose of the aircraft should be maintained in the normal descent attitude throughout the maneuver.

The amount which the nose is swung in the opposite direction from the bank should be such that the *original ground track is maintained*. Recovery is accomplished by raising the low wing and simultaneously easing rudder pressure as the wings are leveled and the pitch attitude adjusted to the normal glide attitude. If the control force on the rudder is removed abruptly, the nose will swing too quickly into line and the plane will tend to acquire excess speed.

SIDE SLIP

In a side slip, the longitudinal axis of the airplane remains on the same heading throughout the maneuver, but the flight path has a sideways component. The airplane is banked in the direction in which the slip is intended and then prevented from turning by the application of *opposite* rudder. As long as these control pressures are held, the airplane will maintain a straight flight path several degrees to the same side of the nose as the low wing. This maneuver also decreases the forward travel while increasing the rate of descent. Further clarification will result if the forward and side slips, as

FLIGHT PATH

Fig. 3-35. Forward Slip

shown in figures 3-34 and 3-36, are compared.

To maintain a constant heading and a straight flight path while slipping, it is necessary to have the aileron and rudder control pressures properly balanced. Too much aileron or too little opposite rudder will cause a turn in the direction of bank. In contrast, too much rudder with too little opposite aileron will cause a yaw away from the bank. To steepen the descent, both aileron and rudder must be increased in a coordinated, proportionate manner.

ACCEPTABLE PERFORMANCE IN SLIPS

The aircraft should be placed in either the forward- or side-slip configuration smoothly, without jerky or erratic control usage. The aircraft should maintain the desired track across the ground with coordinated control changes to compensate for drift. The return to normal glide attitude should be accomplished smoothly and with coordinated control usage.

Fig. 3-36. Side Slip

CHAPTER 4

ADVANCED MANEUVERS

SECTION A – SLOW FLIGHT

During flight training, maneuvers will be introduced that cover the full scope of an aircraft's handling characteristics and responses to control movements. The general purpose of these maneuvers is to teach the student how the aircraft responds and reacts in a variety of power, altitude, flap, and gear configurations. Practice of these maneuvers will develop the student's perception, his feel for the aircraft, and his "air" sense.

Slow flight, sometimes called minimum controllable airspeed, is one of the maneuvers that will help the new student develop feel for the controls and dramatically relate *altitude and airspeed control.* The objective of practicing slow flight is to learn the relationship of power to altitude control and elevator control to airspeed, to assess the reduction of control effectiveness during slow flight, and to determine the power required for flight at low airspeeds. Practice of slow flight prepares the student for the maneuvering techniques and the positive control necessary at the speeds used *immediately after takeoff and on final approach just before touchdown.*

SLOW FLIGHT PROCEDURES

Slow flight, or minimum controllable airspeed, is practiced at an airspeed above flaps-up stall speed, as shown in figure 4-1, but sufficiently slow so any reduction in speed or an increase in load factor will result in immediate indications of an imminent stall.

Aircraft of recent design are required to have positive stall warnings that begin between 5 and 10 miles per hour above stall speed. According to the low speed end of the green arc of the airspeed indicator in figure 4-1, this speed is 63 m.p.h. An acceptable speed to use for slow flight practice is about 10 m.p.h. above the stall speed listed in the aircraft owner's manual or, as shown in figure 4-1, 73 m.p.h.

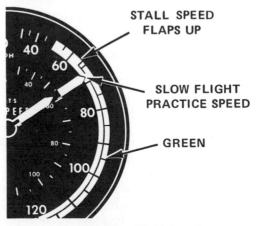

Fig. 4-1. Slow Flight Speed

Fig. 4-2. Transition From Level Flight To Slow Flight Attitude

The task of entering slow flight from cruise flight will develop smoothness and coordination of elevator, rudder, and power controls. The initial problem is to perform a smooth transition from level–flight attitude to slow-flight attitude while maintaining altitude, as shown in figure 4-2.

The instructor will demonstrate slow flight the first time and then will set the nose and wing attitude for straight-and–level slow flight. He will also set the power required to maintain a constant altitude. Figure 4-3 shows how the nose attitude might look from inside the aircraft using visual and instrument references.

Fig. 4-3. Nose Attitude For Slow Flight

The wingtips can also be used as a visual attitude reference. In figure 4-4, the position and angle of the wingtips in relation to the horizon indicate to the pilot that the wings are level and the nose is high.

The procedure used to enter slow flight is: *first*, pull on carburetor heat (if normally used when power is reduced); *then*, reduce power a few hundred r.p.m. (or a few inches of manifold pressure) below that required to maintain altitude in slow flight. The student should then

increase back pressure on the control wheel to *reduce* airspeed.

Fig. 4-4. Wingtip Attitude In Slow Flight

While the airspeed is decreasing, the student should increase back pressure on the control wheel just fast enough to maintain altitude. If the elevator pressure is applied too fast, the airplane will climb; if the application is too slow, altitude will be lost. During this transition, the trim tab should be used to remove control pressure.

When the aircraft nears the desired attitude and airspeed, the power must be increased to a setting which will maintain altitude in slow flight. Small power adjustments then are made to maintain a constant altitude and the aircraft retrimmed as necessary to remove control pressures.

As the aircraft slows, the pilot also will notice that the rudder pressure requirements change. *Right rudder pressure* becomes necessary for proper coordination as the aircraft slows and enters this high angle of attack, high torque, and pronounced P-factor flight condition. The pilot should refer to the support instrument (the ball of the turn coordinator) to assist in determining the amount of rudder required. A feeling of pressure pushing against the student's body to the right also will tell him that right rudder is needed. (See Fig. 4-5.)

Fig. 4-5. Right Rudder Required During Slow Flight

At the airspeed used for slow flight, there will be a sensation of inefficient controls. The controls will feel "mushy" and it will be necessary to use greater control movements than normally required for the same attitude displacement. The aircraft's response to the control movements also will be slow.

ALTITUDE AND AIRSPEED CONTROL

When performing slow flight at minimum controllable airspeed, the *primary* function of the flight controls becomes very apparent. An attempt to climb by simply raising the nose with the elevators will not be successful; rather, the airspeed will decrease, immediate indications of an approaching stall will occur, and the altitude will decrease. Attempts to lose altitude by lowering the nose with the elevators will cause a decrease in altitude, but the *airspeed will also increase beyond acceptable limits.*

The *correct procedure to gain altitude is to apply power.* A very small increase in pitch attitude may be necessary to maintain airspeed. *To lose altitude,* power *must be reduced* in conjunction with a small reduction in pitch attitude. This relationship of power as a *primary altitude control* and the use of elevators as a *primary airspeed control* can be further clarified by an in-flight demonstration.

In this demonstration, the student is given the throttle and the instructor takes charge of the elevator control. The student is instructed that he can use any power setting he wants and change it as often as he desires. The instructor will then maintain a given airspeed, such as 75 m.p.h., regardless of where the power is set.

If the student applies full power, the instructor simply increases elevator pressure to adjust the nose position to an attitude which produces the desired airspeed. This is really nothing more than a full power, 75 m.p.h. climb.

If the student reduces power to idle, the elevator controls are used to lower the nose attitude to produce the desired airspeed. This can be accomplished by establishing a 75 m.p.h. power-off descent. This exercise should graphically clarify that the *airspeed is controlled primarily with elevator pressure.*

This demonstration should also point out that *with a constant airspeed, it is power that controls altitude.* With the instructor maintaining a constant airspeed, the student can control the altitude. If the aircraft is climbing, a slight reduction of power can be made to remain in level flight. Further reductions in power will cause a descent. (See Fig. 4-6.)

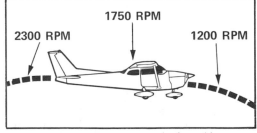

Fig. 4-6. Power Controls Altitude

The relationship between airspeed and elevators is a fixed relationship and can be shown by another simple in-flight demonstration. First, the aircraft should be trimmed so the airspeed is maintained at 75 m.p.h. and the power set so the *altitude* remains constant.

Next, the position of the trim tab should be marked. Marking the trim tab, in effect, marks the elevator position since the trim is attached directly to the elevator through pulleys and cables. Also, the position of the tachometer needle and manifold pressure needle (if installed) should be marked on their respective indicators. Now, any time the pilot wishes to go to 75 m.p.h. in level flight, all he needs to do is set the power at the mark on the instrument and set the marker on the trim tab.

By flying with the hands essentially off the controls and using the ailerons only to keep the wings level, the pilot simply reduces power to the marks on the instrument and sets the trim tab to its pre-established mark. The aircraft may oscillate slightly, but if the pilot allows the aircraft to stabilize, it will *return* to 75 m.p.h. and *maintain* altitude.

The purpose of this exercise is not a recommendation to place marks on the indicators for each condition of flight, but to *fix* in the student's mind that *there are direct fixed relationships*. During practice, the student should observe and remember attitudes, power settings, and speeds so he easily can visualize and duplicate the maneuver day after day.

Emphasis will be placed on the ability to establish known values of attitude and power and predict the outcome in terms of airspeed and altitude because of their importance when practicing in the traffic pattern later.

As an example, if the aircraft is low and a little short of the runway on final approach, it is very important that the pilot add power to extend his path, as shown in figure 4-7. If the *nose is pulled up*, the glide path will be *steeper* and the aircraft will end up still *shorter of the runway*.

CLIMBS, DESCENTS, AND TURNS IN SLOW FLIGHT

Climbs and descents in slow flight will be practiced. As previously discussed, altitude is controlled by power. To climb, the pilot adds power; to descend, he reduces power.

In slow flight, turns will also be practiced. Again, as in any turn, part of the total lift force is *diverted* to make the aircraft turn.

In cruising flight, the pilot pulls the nose up a little and loses a little speed to gain the extra vertical lift component needed to counteract gravity; however, slow flight is just above stall warning indications, so the loss of airspeed will bring the aircraft closer to the stall. The steeper the turn, the closer the aircraft will be to the stall; so *power*, rather than additional elevator back pressure, is added to increase the vertical component of total lift and maintain altitude.

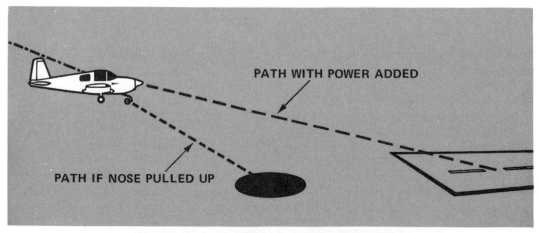

PATH WITH POWER ADDED

PATH IF NOSE PULLED UP

Fig. 4-7. To Extend Glide, Power Should Be Added

WING FLAPS

Slow flight practice also will include using extended flaps and landing gear, if the aircraft in use has retractable gear. During practice, the pilot is instructed to lower 10°, or one notch, of flaps at a time. At each position, he adjusts the attitude to maintain airspeed, adjusts power to maintain altitude, and adjusts trim to relieve control pressures.

It is commonly found that there is very little nose position, power, or trim change with the addition of flaps to the first position (10°-15°); however, with flaps in the second position (about 20°–25°), the nose position must be lowered. An increase in power and change of trim also will be required. When full flaps are lowered (about 40°), the change usually will be quite pronounced. In sequence, the nose position in the aircraft might look like that shown in figure 4-8.

The attitude gyro indications associated with increasing flap deflections may look like those shown in figure 4-9. If it is found that the required power settings to maintain level flight at 75 m.p.h. are 1,750 r.p.m. for slow flight with no

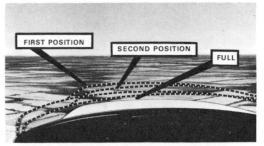

*Fig. 4-8. Nose Position In Slow Flight
With Flaps*

flaps, 1,850 r.p.m. for the first flap position, 2,000 r.p.m. for the second position of flaps, and 2,300 r.p.m. for full flaps, it should be evident that the use of flaps in slow flight requires added power to compensate for the increased drag.

With full flaps, climb capability is very limited. This is a high lift and high drag condition and requires a large percentage of available power simply to maintain altitude. Full power will probably produce only a very slow rate of climb, if any.

Carburetor heat should be in the COLD position in this flight condition as the reduced power output when carburetor heat is ON results in little or no climb capability.

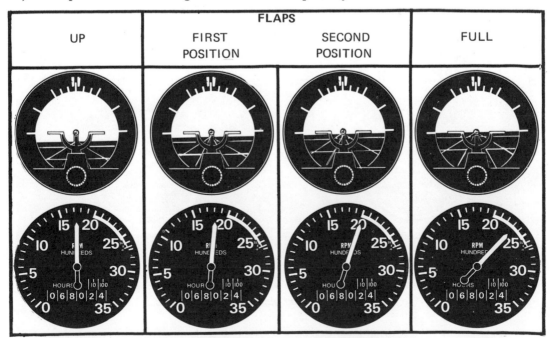

*Fig. 4-9. Typical Pitch Attitude And Power Changes
Required To Maintain Altitude As Flaps Are Added*

The pilot must be aware of the *high drag produced* when he is in a full-flap, slow speed landing configuration and has to abandon the approach and *go around* at the last instant. In that situation, he must add full power and maintain altitude. Attempting to initiate a climb by pulling up the nose before slowly reducing the large flap extension setting may only result in the aircraft losing the limited climb capacity it has and settle to the runway or inadvertently stall.

RETURN TO CRUISE FLIGHT

To return to straight-and-level flight from slow flight, full power is applied. The flaps are raised *slowly*, one position at a time. Back pressure must be added to compensate for the loss of lift as flaps are retracted. The pilot should *hesitate* between flap positions and readjust the trim tab to remove control pressures.

Caution should be exercised when raising flaps. They should be raised slowly, allowing the speed to increase gradually as flaps are retracted. For example, while the aircraft may have a safe margin of airspeed with flaps, as seen on the left airspeed indicator in figure 4-10, sudden or complete retraction of the flaps at that speed may place the aircraft near the stall speed in a no-flap condition as seen in the airspeed indication on the right.

ACCEPTABLE PERFORMANCE FOR SLOW FLIGHT

Acceptable performance is characterized by prompt recognition of the slow flight attitude and the relationship of attitude to airspeed control. The student should also be able to apply power to climb, descend, or maintain altitude at an assigned airspeed. The airspeed should be maintained within five knots, altitude within 100 feet, and heading within 10° of assigned values. Primary emphasis is on airspeed control; however, inadequate area surveillance or an unintentional stall is disqualifying.

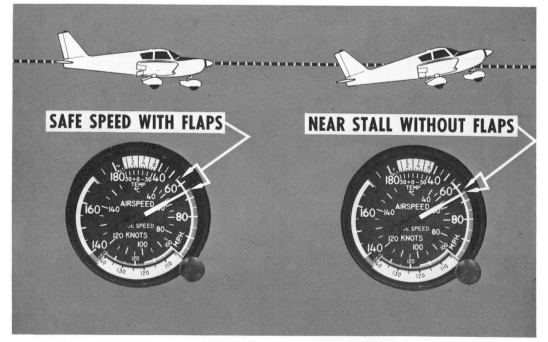

Fig. 4-10. Stall Speed Increases Without Flaps

SECTION B — APPROACH-TO-LANDING STALLS

INTRODUCTION

Within the logical progression of teaching flying skills, stalls are among the next maneuvers practiced in the flight training curriculum. There are a number of misconceptions that should be dispelled before undertaking a detailed explanation of how to stall an aircraft and how to recover from the stall.

Some persons have the erroneous idea that *the engine stops* during the stall and that once in the stall, the pilot is "at the mercy" of the aircraft. Since the only dangerous characteristic regarding stalls is the associated loss of altitude during recovery, the stall is a safe maneuver except when it occurs at such a low altitude that proper recovery cannot be affected before contact with "mother earth."

Although stalls are practiced with an attempt to precisely control airspeed, altitude, attitude, etc., stalls are *not* practiced simply for the sake of being able to stall an airplane with perfection. Stalls are practiced to accomplish two main objectives:

1. To enable the student to sample the stall warnings and handling characteristics of an airplane as it transitions from cruise to slow flight and approaches the stall; therefore, an *awareness* of an imminent stall is a desired objective.

2. If, because of the distractions of the flight conditions (for example, turbulence, inattention, etc.), the pilot should stall the aircraft, he should be able to *promptly and effectively recover from the stall with a minimum loss of altitude.*

Since modern aircraft receive the benefit of advanced aerodynamic knowledge and

design criteria, and *must be able to recover from a stall by itself*, the pilot should approach the practice of stalls with little or no apprehension; however, a "few small butterflies" may be expected and accepted as normal.

CERTIFICATION REGULATIONS REGARDING AIRPLANE STALL CHARACTERISTICS

Before practicing approach-to-landing stalls and the more advanced stall maneuvers covered in chapter 5, the student can gain some insight into what to expect by reviewing a portion of the Federal Aviation Regulations. Part 23 of the Regulations prescribes the airworthiness standards for small airplanes used in flight training. This part contains the regulations pertaining to flight performance and handling characteristics, structure, powerplant, equipment, and the operating limitations.

Before a manufacturer can offer an airplane for sale, it must meet the specifications of this Regulation and this includes flight testing and stall demonstrations. The Regulations *require* that:

1. The pilot can *correct* a roll or yaw up to the stall. This means that the pilot must have effective use of the controls up to actual stall occurrence.
2. The pilot can *prevent* more than 15° of roll or 15° of yaw by the normal use of controls during the recovery from a stall. In short, the controls must be effective during the recovery.
3. There must be a *clear and distinct stall warning* with the landing gear and flaps in any position the pilot choses and in either straight or turning flight. The stall warning must begin between 5 and 10 miles

per hour before the stall occurs, continue into the stall, and end with the recovery. The acceptable stall warning can be buffeting, such as a general shaking or vibration of the aircraft, or a visual or aural stall warning instrument.

The Regulations also state that if the airplane loses more than 100 feet or pitches more than 30° nose-down in the stall demonstration, the information must be listed in the aircraft flight manual. The airplane must perform satisfactorily and meet the specifications described before it is made available to the public.

CAUSES OF THE STALL

Prior to performing the stall maneuver, the factors which cause a stall should be reviewed. Simply stated, a stall is caused by *an excessive angle of attack* which results in the smooth flow of air over the wing breaking away from the wing's upper surface.

The angle of attack is the angle between the wing chord line and the direction that the wing is moving. Air moves toward the wing from the direction in which the wing is moving, or along the flight path of the aircraft. This air, or wind, is called the *relative wind.* The common definition of angle of attack,

and the one the student should recall from earlier ground training, is that the *angle of attack is the angle between the chord line and the relative wind.* (See Fig. 4-11.)

The wing can move through the air at several different angles with respect to the relative wind, as shown in the diagrams of figure 4-12. It is important to remember that the relative wind, shown by the arrows, is *seldom* parallel with the horizon and that the angle of attack is measured between the chord line and relative wind (not the horizon). This means it is possible to have a high angle of attack with the nose low and a low angle of attack with the nose high.

In normal flight, the air flows over the top of the wing and "hugs" the wing; but even in cruise, at a certain point back on the wing, the smooth air eventually tears away from the wing. This is called the *separation point.*

As the angle of attack increases, the separation point *moves forward,* as shown in figure 4-13. As the separation point moves forward because of the increasing angle of attack, it eventually reaches a point where too much air has separated from the wing. There is no longer sufficient lift to support the air-

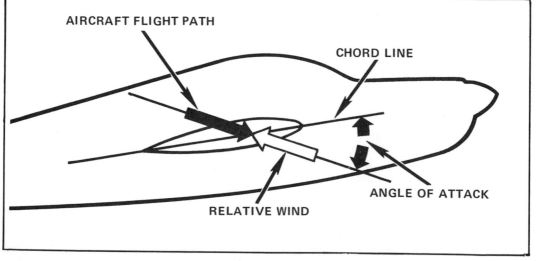

Fig. 4-11. Angle Of Attack

craft and the resulting drag has increased greatly. This point is called the stall. It should be noted that lift isn't all lost; however, not enough lift remains to support the aircraft.

To recover from the stall, the pilot reduces the angle of attack and restores smooth airflow over a greater portion of the wing producing sufficient lift to support the aircraft.

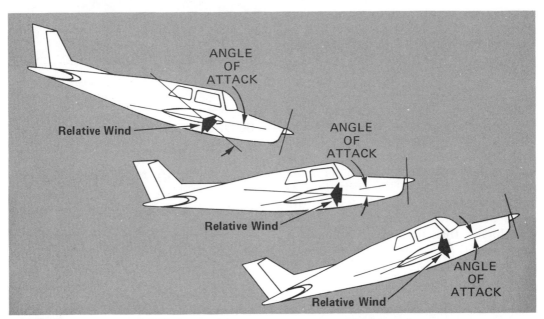

Fig. 4-12. Angle Of Attack Not Related To Nose Position

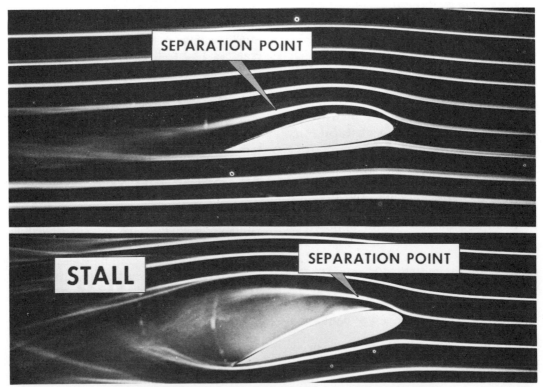

Fig. 4-13. More Air Separates As Angle Of Attack Is Increased

FACTORS AFFECTING STALL SPEED

On the airspeed indicator, the calibrated stall speed in *cruise configuration* may be read at the *low speed end* of the green arc. The calibrated stall speed with flaps down (in landing configuration) may be read at the low speed end of the white arc. (See Fig. 4-14.)

Fig. 4-14. Calibrated Stall Speeds Shown On Airspeed Indicator

The stall speeds noted on the airspeed indicator are used as guide numbers only, since the smooth airflow separation point that causes a stall can be made to occur at faster airspeeds if the wing loading is increased. For example, *if the load is increased* from the normal one G to two Gs, the stall separation point will occur at a speed approximately 40 percent higher than the normal stall speed. Increasing the load from one G to two Gs can be done by a sudden application of elevator pressure to raise the nose. *Exceeding the aircraft gross weight* by overloading with fuel, passengers, and baggage also increases the wing loading and *increases the stall speed.*

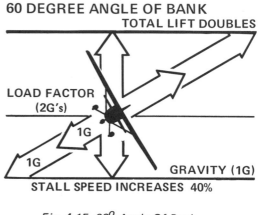

Fig. 4-15. 60° Angle Of Bank Doubles Load Factor

Another way of increasing the effective load on the aircraft is to bank the aircraft steeply while maintaining altitude. In straight-and-level flight, the load on the aircraft is equal to the force of gravity. In a turn, as total lift increases to support the aircraft, the centrifugal force and the effective load on the aircraft increase.

At a *60° angle of bank*, the *total lift* required to support the aircraft *is doubled* and the *load factor* imposed on the aircraft *is doubled.* (See Fig. 4-15.) At this angle of bank, the stall speed increases approximately 40 percent. For example, an aircraft that stalls at 63 m.p.h. in straight, unaccelerated flight will stall at 88 m.p.h. in a 60° bank.

It should be clear that the stall speeds shown on the airspeed indicator are guidelines for level, unaccelerated flight only. When the aircraft turns, the pilot must recognize that the stall speed increases. The stall speed increase is very small at shallow angles of bank; however, at a 45° angle of bank, the stall speed in most training aircraft will probably increase about 14 m.p.h. At angles higher than 45°, the stall speed increases very rapidly.

PRACTICE AREAS

Stalls will be practiced in an area previously designated by the instructor as the local practice area. As a general guideline, before beginning stall practice, the student should attain an altitude of *at least 2,000 feet above ground level* (AGL) This is an altitude that will generally permit stall recovery at least 1,500 feet above the ground.

During stall practice, as well as in other maneuvers, the area must be "cleared" of other aircraft. Clearing the area can be done in a number of ways: by making two 90° turns, two 180° turns, or a 360° turn; however, the number of turns and in which direction really aren't the important points. The *object* of the turns is to *clear the area.* The pilot should look over the entire area, especially at the flight altitude. He should be sure that there are no other aircraft in the area, and if there are aircraft in close proximity, the pilot should wait until they are well clear before performing the maneuver.

Visibility is a factor in how thoroughly and how often the pilot clears the area. Visibility of three miles is the absolute minimum at which stalls should be practiced in uncongested areas; however, visibility of five miles will provide a more comfortable margin. In addition, Federal Aviation Regulations require that an aircraft below 10,000 feet MSL remain 500 feet below, 1,000 feet above, and 2,000 feet horizontally from any clouds.

INTRODUCTION TO THE APPROACH-TO-LANDING STALL MANEUVERS

After the area has been cleared and a satisfactory altitude established, the first introduction to a stall usually will begin from a power-off, wings-level glide, such as that used when landing the aircraft. This is called an approach-to-landing stall. (See Fig. 4-16.)

To enter the stall, the pilot will apply carburetor heat (if required), reduce power to idle, establish a normal glide, apply full flaps, and trim the aircraft. Next, he will begin to slow the aircraft by applying back pressure to the elevators.

As the airspeed slows, an objective is to gain a "feel" for the control pressures and responses as the aircraft approaches the stall. One systematic method the student can use during this demonstration is each time he loses five miles per hour of airspeed, he should *hesitate* at that airspeed for a moment and *move the aileron and rudder controls* back and forth to assess the control responsiveness. As he slows another five miles per hour, he should repeat the control feel analysis until the aircraft produces stall warnings.

As speed slows, the response of the aircraft to control pressures becomes slower, and greater displacement of the controls is necessary to achieve the desired results. The feeling is sometimes referred to as "mushy" or "soft." Compared to the more solid feel of the controls at cruise speed, the control wheel feels more like it has a "rubber connection" to the control surfaces. Heavier control pressures are required to maintain the desired pitch attitude. These are clues that the aircraft is slowing and approaching a stall.

As the stall is approached, there are other clues that inform the pilot the aircraft is slowing. *Sound* is one of those clues. As learned during slow flight practice, the tone and intensity of the slipstream noise as well as changing engine sounds provide useful clues.

As the aircraft approaches the stall, the student may also get a feeling that the airplane is "mushing." Sometimes this is referred to as a feeling similar to sitting on sponge rubber, or a slight sinking sensation. The student should not be surprised if these "seat of the pants"

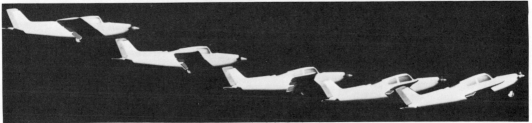

Fig. 4-16. Entering The Approach-To-Landing Stall (Power-Off Stall).

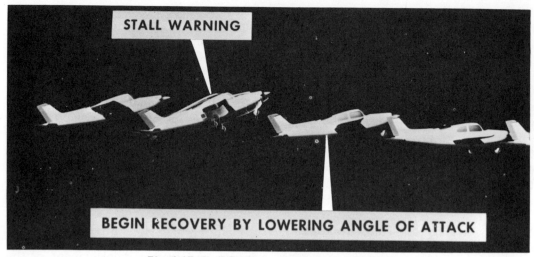

STALL WARNING

BEGIN RECOVERY BY LOWERING ANGLE OF ATTACK

Fig. 4-17. Stall Recovery At Beginning Of Stall

feelings aren't very noticeable during the first demonstration. The feeling develops and becomes more noticeable as experience is gained.

The stall warnings will begin 5 to 10 miles per hour before the stall. The warning may be a buzzer, horn, light, buffeting of the controls, vibrations, miscellaneous noises, or any combination of these; the student must learn to recognize those applicable to his aircraft.

As the aircraft continues to slow down, the stall indications become more noticeable. One of the purposes of stall practice is *to recognize these indications.*

In the first demonstration, the student will recover to straight-and-level flight after observing the stall warnings and *before* the stall is fully developed. Coming close to the stall, but not fully stalling, is referred to either as an *imminent stall or an approach to a stall.* To recover, the angle of attack must be reduced, as shown in figure 4-17.

The pilot's *first action* is to lower the nose; simply *releasing* the back pressure is usually sufficient. The nose should be lowered to level flight attitude, or very slightly below, and the pilot should use whatever control pressures are necessary to accomplish this attitude change.

*Fig. 4-18. Visual And Instrument References
Just Before Stall Occurs*

The visual and instrument references just before the stall occurs are shown in figure 4-18. Figure 4-19 depicts the visual and instrument references as back pressure is released and the angle of attack is reduced.

*Fig. 4-19. Visual And Instrument References
At Beginning Of Stall Recovery*

Power is applied *simultaneously* as back pressure is released. The throttle should be moved promptly and smoothly to obtain all available power and right rudder pressure added to compensate for the torque or turning effect. Care should be

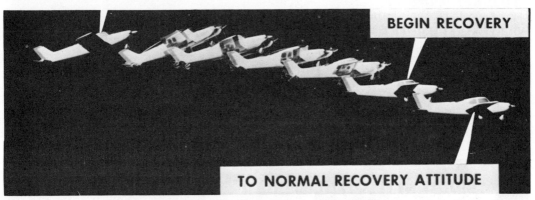

BEGIN RECOVERY

TO NORMAL RECOVERY ATTITUDE

Fig. 4-20. Full Stalls (Delayed Recovery)

taken not to "jam" the throttle forward since this may cause the engine to falter. As power is added, the carburetor heat control should be moved to the COLD position.

Straight-and-level flight at cruise airspeed with a *minimum loss of altitude* is desired in stall recovery. Coordinated, normal use of ailerons and rudders is used to achieve this result. As airspeed increases following recovery, less right rudder pressure is required and power can be reduced to normal cruise power setting.

In the event that the aircraft's nose attitude is not lowered enough during the recovery, there is a possibility of entering a *secondary stall*. The pilot should use *smooth but positive* pressures during recovery. On the other hand, an excessive nose-down recovery attitude will result in too much altitude loss and an unacceptably high r.p.m. and air speed.

Stall recovery should be followed by climbout in climb configuration to an altitude of *at least 300 feet above the altitude at which full control effectiveness is regained.*

FULL STALLS

After practicing approaches to stalls, the student will be instructed in the technique commonly called a "full stall" or a "delayed recovery stall." The first part of full-stall practice requires exactly the same procedures as the approach to a stall. The warnings of the approach to a stall are the same, but the difference in full-stall practice is that the angle of attack is increased *beyond* the point where recovery would normally be started in the approach-to-a-stall maneuver and the aircraft is *forced* "deeper" into the stall before recovery begins. (See Fig. 4-20.) In the *full stall*, recovery is started *after* the nose pitches down, as shown in figure 4-20. The pilot decreases the angle of attack by releasing back pressure to lower the nose. The point at which back pressure is released and the approximate pitch attitude during recovery is shown in figure 4-21. As this happens, the student may get a sensation in his stomach similar to that experienced when an elevator starts downward or in a car going over a hill on a road where the occupant is momentarily light in the seat. This sensation is of short duration and disappears as soon as the stall recovery is completed.

As the airspeed begins to build following the recovery, the pilot gently and smoothly readjusts the pitch attitude. Again, if the nose is pulled up too rapidly following recovery, a secondary stall may result. On the other hand, a pullup too late may result in excessive speed. (See Fig. 4-22.)

During full-stall practice, it is possible that the aircraft will tend to turn and roll to one side as the nose pitches. (See

Fig. 4-23.) If that happens, the pilot simply uses *coordinated aileron and rudder pressures* to level the wings at the same time he is applying power and re-establishing the pitch attitude.

The pilot should apply power as he begins the stall recovery. The same guidelines apply to power and throttle control as in approaches to stalls. The application should be smooth and prompt. Again, recovery should be followed by a climb-out in climb configuration to an altitude of at least 300 feet *above* the altitude at which full control effectiveness is regained. After level-off, power may be reduced to the normal cruise setting and the aircraft retrimmed as necessary.

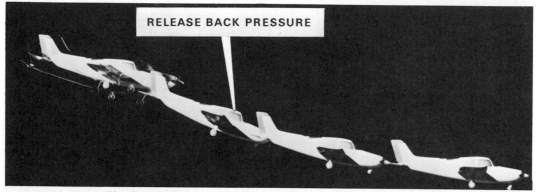

Fig. 4-21. Full-Stall Recovery Begins After Nose Pitches Down

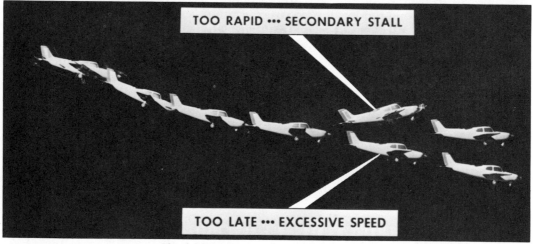

Fig. 4-22. Incorrect Full-Stall Recoveries

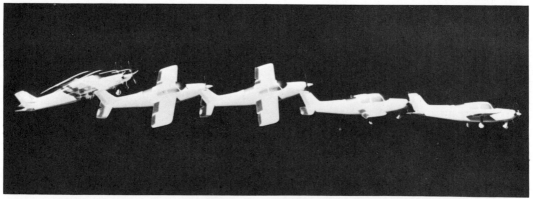

Fig. 4-23. Coordinated Aileron And Rudder Control Is Used In Stall Recovery

Some airplanes may be placed in a flight condition of a high rate of descent with the control wheel pulled all the way back. This is a stabilized, high angle-of-attack condition where the elevators are in the full-up position so the nose can't be raised any higher. If the student encounters this condition while attempting to enter a stall, he should recover, re-enter the stall, and achieve a *higher* nose attitude. The stabilized, high angle-of-attack condition is similar to what is encountered in aircraft with limited elevator travel, or so-called "stall-proof" aircraft. Recovery from this condition should be made in the manner previously described.

ACCEPTABLE PERFORMANCE FOR APPROACH-TO-LANDING STALLS

The student will be judged on his ability to recognize the indications of an imminent stall and take prompt, positive recovery actions. When practicing imminent stalls, the student must not let a full stall develop. During full stall practice, he is expected to immediately recognize when the stall has occurred and take prompt corrective action. The recovery must be performed without excessive airspeed, excessive altitude loss, a secondary stall, or loss of control such as a spin.

SECTION C – STEEP TURNS

After proficiency has been gained in the basic flight maneuvers, the pilot applicant will be introduced to the more advanced maneuvers, including steep turns. Although steep power turns are not required for a private pilot flight check, they have direct application to turns about a point. In addition, the commercial flight test specifically requires consecutive right and left steep power turns using a bank angle of at least 50°. The following discussion applies mainly to 45° banked steep turns, however, the principles apply equally to the steeper turns (50° to 60°) required of the commercial pilot.

TURN SPEEDS

Steep turns are started at the speed recommended by the manufacturer for this maneuver. If no recommendation is listed, cruise speed or maneuvering speed, whichever is lower, is used. The angle of bank is suggested by the instructor before entering the turn and is usually about 45° for initial training.

LIFT FORCE

In a turn, the total lift force is diverted and part of the lift is used to make the aircraft turn, as stated in a previous chapter. When the lift is diverted, the vertical component of lift used to support the aircraft is less than gravity and the aircraft will descend unless some action is taken to increase lift. (See Fig. 4-24.) Total lift must be increased so that sufficient vertical lift is available to counterbalance gravity. This is done by increasing the angle of attack with back pressure applied to the control wheel. This action also results in a decrease in airspeed.

In a 45° banked turn, the additional back pressure required and resulting loss of airspeed is more noticeable than in medium-banked turns. In an aircraft with a fixed-pitch propeller, the decrease

in airspeed is also accompanied by a decrease in r.p.m. with a resulting power loss. Consequently, more airspeed is lost in the attempt to maintain altitude.

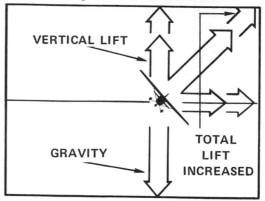

Fig. 4-24. Lift Must Be Increased To Maintain Altitude In A Turn

The problem that arises is that as the *angle of bank is increased*, the *stall speed increases*, and as *bank increases*, *airspeed decreases* as the pilot attempts to maintain altitude. Thus, the stall speed and airplane's airspeed approach the same value.

For example, an aircraft with a fixed-pitch propeller that normally stalls at 63 m.p.h. will stall at 89 m.p.h. in a 60° bank. If the aircraft enters the turn in a 60° bank at its maneuvering speed of 110 m.p.h. and the airspeed drops to approximately 95 m.p.h., the *margin between aircraft speed and stall speed will only be six miles per hour.* (See Fig. 4-25.)

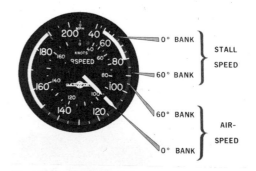

Fig. 4-25. Stall Speed And Aircraft Airspeed Converge In A Steep Turn

STEEP TURN PROCEDURES

When the student understands and can visualize the attitude required in level flight and the attitude required in the turn, the problem of pitch attitude adjustment during the turn entry becomes evident. A smooth transition from level flight into the turn must take place.

TURN ENTRY

Bank should be established at a moderate rate and not rushed. Frequently, the beginner feels that because the angle of bank is to be greater, the roll into the turn must be faster; however, this is not the case. Better results will be obtained with a slower and well-planned entry, since the turn entry is judged on smoothness and coordination.

The normal procedure in an aircraft with a fixed-pitch propeller is to *add power* after the bank is established to maintain r.p.m. and, consequently, maintain altitude. Even with the application of power, there will be an airspeed reduction and a requirement for considerable back pressure. The trim tab can be used to help relieve pressure in the turn, but if it is used, the student must remember that the trim must be changed again as soon as he rolls to level flight.

Normally, in an aircraft with a constant-speed propeller, the power is at cruise setting throughout the maneuver. However, the airspeed will still decrease and control wheel back pressure will be required.

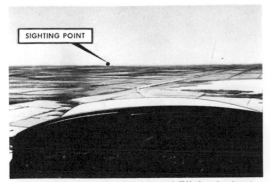

Fig. 4-26. Straight-And-Level Flight Attitude

VISUAL REFERENCES

In figure 4-26, the nose attitude for straight-and-level flight at maneuvering speed is shown. The sighting point above the nose of the aircraft would be approximately at the point of the arrow.

In figure 4-27, the aircraft is established in a 45° banked left turn and is maintaining a constant altitude. The sighting point on the nose is now higher than in straight-and-level flight due to the required back pressure. Much of the instrument panel now appears higher above the horison than in straight-and-level flight.

Fig. 4-27. Attitude In 45° Left Turn

Figure 4-28 shows the aircraft in a 45° banked right turn also maintaining altitude. The most apparent difference is that in a right turn, the nose *seems to be considerably lower* than in the left turn; however, the sighting point over the nose of the aircraft is *in the same position* as in the left turn.

Fig. 4-28. Attitude In 45° Right Turn

INSTRUMENT REFERENCES

In a left 45° banked turn using instrument references, the bank index, pointed out by the arrow at the top of figure 4-29, shows the number of degrees of bank. It should be noted that the nose of the little airplane is slightly above the horizon line indicating a nose-high attitude.

BANK INDEX

*Fig. 4-29. Instrument References In 45°
Left Turn*

In a 45° right banked turn, there is no change in the attitude references except that the artificial horizon shows a bank to the right. The nose attitude is *identical* to the left turn indication. (See Fig. 4-30.)

*Fig. 4-30. Instrument References In 45°
Right Turn*

LEVEL FLIGHT

45° LEFT TURN

*Fig. 4-31. Elevator Back Pressure Required To
Maintain Altitude*

TURN ROLL OUT

The procedures for rolling out of a steep turn are the reverse of those used during the roll into the turn. As the bank decreases, the back pressure must be decreased gradually to avoid gaining altitude. (See Fig. 4-31.)

Fig. 4-32. Attitude Changes When Rolling From Left To Right Steep Turn

When some proficiency has been gained the student will be asked to turn 720° in one direction and then 720° in the other direction with no hesitation between the turns. The student simply rolls directly from one bank into the other. (See Fig. 4-32.) This means that back pressure must be reduced during rollout and then reapplied as the angle of bank increases in the opposite direction. Power changes must also be coordinated with the pitch and bank attitude changes.

OVERBANKING TENDENCY

The student also will notice an apparent necessity to *cross-control* in a left turn to maintain a constant angle of bank and coordinated flight. This is caused by the overbanking tendency and is most pronounced at high angles of bank.

The *overbanking tendency* is caused by the fact that when an aircraft is in a steep turn, the wing on the outside of the turn travels farther than the inside wing. While traveling farther, the outside wing has a faster airflow over its surface. The higher airflow speed results in *more lift on the outside wing.* (See Fig. 4-33.)

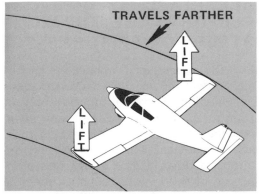

TRAVELS FARTHER

Fig. 4-33. Overbanking Tendency

This increased lift tends to make the aircraft roll steeper into the turn. To counteract this tendency, slight aileron pressure *opposite* to the direction of turn must be applied to keep the aircraft from over banking.

At the same time, rudder pressure is required *throughout* the turn. The tail section, *because of its distance aft the center of gravity of the aircraft, does not track in the same arc.* Rudder pressure is needed to streamline the fuselage in the arc of the steep turn. The force needed to streamline the fuselage properly is greater than the amount required to counteract the aileron drag on the high wing, as discussed under adverse yaw. (See Fig. 4-34.) The use of cross-control technique keeps the aircraft from overbanking and maintains coordinated flight even though everything the student has learned to this point seems to contradict the statement.

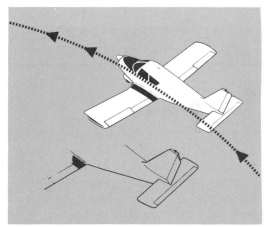

Fig. 4-34. Rudder Needed Throughout Steep Turn

TORQUE AND P-FACTOR

One other slight difference noted in a steep turn is that *less* overbanking tendency is evident in a right turn than in a left turn. Torque and P-factor which tend to roll the aircraft to the left work *against* the overbanking tendency in a right turn, but *increase* the tendency when turning to the left. (See Fig. 4-35.)

Steep turns are like any other maneuver in that a continuous series of small adjustments must be made as the turn progresses. Corrections during steep turns follow the same general rules given before. When the aircraft is deviating from the desired attitude or altitude, the student should use the *two-step* method: first, stop the deviation; and second, make a small correction back to the desired settings.

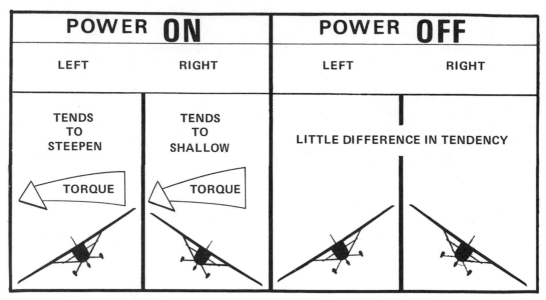

POWER **ON**		POWER **OFF**	
LEFT	RIGHT	LEFT	RIGHT
TENDS TO STEEPEN	TENDS TO SHALLOW	LITTLE DIFFERENCE IN TENDENCY	
TORQUE	TORQUE		

Fig. 4-35. Torque Effects Overbanking Tendency In Steep Turn

ALTITUDE CONTROL

In steep turns of 45° or more, altitude corrections deserve special consideration. If altitude is lost because of a nose-low attitude, simply pulling back on the control wheel *is not* a satisfactory correction. The force created by back pressure on the control wheel in a 45° bank raises the nose, but also tightens the turn *just as much* as it lifts the nose. This procedure simply increases the load on the aircraft and does little to correct for altitude. The proper response is to temporarily *reduce the angle of bank* slightly and the nose simultaneously will rise. When this correction has been made, the pilot then should return to the desired angle of bank and apply slightly more back pressure than previously held.

The rollout procedures for steep turns are performed in the same manner as those for gentler banked turns; however, the rollout should begin approximately 20° *before* reaching the desired rollout heading. The perceptive student pilot may notice a difference in the control pressures required during the roll into a steep turn and those required during the rollout. Generally speaking, *less* rudder and aileron pressure is required during the roll into the turn than during the rollout. This is because the control pressures exerted during the rollout must *overcome* the aircraft overbanking tendency which is present during the steep turn.

ACCEPTABLE PERFORMANCE FOR 720° STEEP POWER TURNS

For the commercial applicant, performance will be evaluated on the basis of planning, coordination, smoothness, prompt stabilization of the turns, maintenance of constant bank and altitude, and orientation. The ability to roll from one turn directly into a turn in the opposite direction will demonstrate the advanced coordination skills desired in this maneuver. The angle of bank should be at least 50° and turn entries and recoveries should be accomplished promptly and smoothly, with appropriate power adjustments. The entry altitude must be maintained within 100 feet and bank must not vary more than 5°. Slips or skids should be immediately corrected and recovery headings should be within 10° of the entry headings.

SECTION A—TAKEOFF-AND-DEPARTURE AND ACCELERATED STALLS

In chapter 4, the student was introduced to the power-off stall maneuvers, commonly called approach-to-landing stalls. Once proficiency has been gained in these power-off stall maneuvers, demonstration and practice of approaches to stalls and full stalls *with power* is given.

An analysis of FAA accident statistics indicates that the performance of inadvertent stalls often results from lack of knowledge regarding the aircraft's handling characteristics during the takeoff and landing phases of flight operation. Therefore, the objectives of practicing stalls and stall recoveries are to determine that the pilot can *promptly recognize* the control responses, kinesthetic sensations, and feel of the aircraft as the aircraft *approaches* a stall. A further objective is to enable him to make prompt, effective recoveries from both partial and *complete* stalls encountered in all normally anticipated flight situations.

TAKEOFF- AND - DEPARTURE STALLS

Takeoff-and-departure stalls are normally practiced in a series of three: a straight-ahead stall, a turning stall to the left, and a turning stall to the right. This presentation will separate the discussion of straight-ahead stalls from turning stalls to simplify the explanation. The student can combine the techniques he will learn in straight-ahead stalls with turns in order to perform the so-called departure stalls.

STRAIGHT-AHEAD STALL

A straight-ahead stall may be encountered during takeoff if the pilot attempts to lift the aircraft from the runway too soon and applies excessive back pressure on the control wheel, thereby producing an extremely nose-high attitude and high angle of attack. It might also occur when a pilot is flying at a low altitude over terrain which increases in elevation faster than the aircraft is able to climb.

The recommended climb power setting, as specified in the aircraft owner's manual, is used for stall practice, and stalls are generally performed at altitudes of at least 2,000 feet above ground level (AGL).

It should be clear that if the airspeed is near cruise, the power set for climb, and elevator back pressure applied, the aircraft will be in a rather extreme nose-high attitude by the time it reaches stall speed. (See Fig. 5-1.) Therefore, while maneuvering the aircraft to "*clear the*

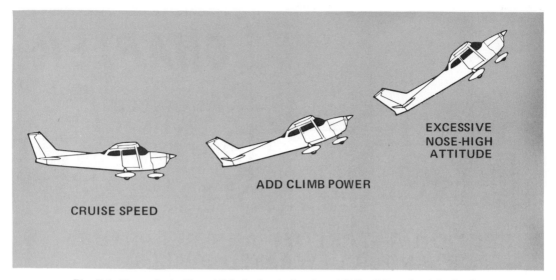

EXCESSIVE
NOSE-HIGH
ATTITUDE

ADD CLIMB POWER

CRUISE SPEED

Fig. 5-1. Excessively Nose-High Attitude Produced If Stall Practiced At Cruise

area," it is recommended that the student reduce power and dissipate excess airspeed while maintaining a constant altitude. After the student rolls out on the heading desired for the maneuver, he is ready to practice the stall.

In order to prevent an abnormally high pitch attitude before the stall, it is important to *initiate the climb at liftoff speed* and increase the angle of attack *progressively* until the stall occurs. As liftoff speed is attained and back pressure added to increase the angle of attack, power is simultaneously increased to the recommended climb power setting. In order to control the aircraft in the high pitch attitude encountered, the student must not attempt to look *over* the nose, but rather *alongside* the nose.

CLUES TO THE APPROACHING STALL

As the airspeed is reduced, the student should develop an awareness of the various clues to the approaching stall. For example, the noise of the slipstream rushing over the aircraft will decrease. However, engine noise will continue at a rather high level. The engine frequently takes on the sound of "laboring" under a load, somewhat like a car "lugging" up a steep hill.

The aileron, rudder, and elevator controls develop an increasing mushiness or sluggish response. In contrast to the situation encountered in approach-to-landing stalls, the propeller slipstream causes the elevator and rudder controls to be more effective and responsive than in the approach-to-landing stalls. Many instructors recommend that as each five mile per hour loss of airspeed occurs, the student move the ailerons, rudders, and elevator so he can gain a feel for the changing response and new relationships of control pressures. At the high power settings and increasingly high angle of attack, there is more torque and P-factor, so additional right rudder pressure is required.

The nose attitude in the straight-ahead takeoff and departure stall should resemble the illustration in figure 5-2. The visual relationship of the nose and horizon are shown on the left and the instrument indications are pictured on the right. The angle which the wingtips make with the horizon can also provide a useful indication as to the aircraft's pitch attitude, as seen in figure 5-3.

RECOVERY PROCEDURES

The procedures for recovery from an approaching stall are the same as detailed

Fig. 5-2. References In Straight-Ahead Takeoff And Departure Stall

Fig. 5-3. Angle of Wingtips With Horizon
In Takeoff and Departure Stall

in chapter 4. The pilot should *decrease the angle of attack* by lowering the nose to the level flight attitude or slightly below. Simultaneously, *full available power should be applied* (if not already at full power) and coordinated aileron and rudder pressures used to return the aircraft to straight-and-level flight, as shown in figure 5-4.

FULL STALLS

After the student is familiar with the straight-ahead *approach to a stall*, the stall warning clues, and has a feel for the

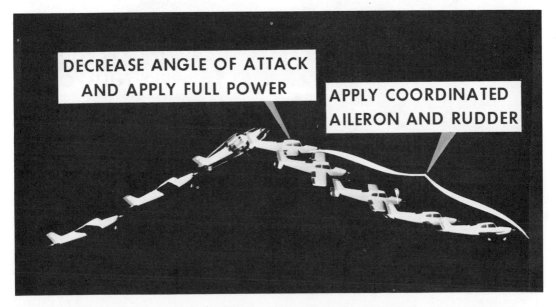

DECREASE ANGLE OF ATTACK AND APPLY FULL POWER

APPLY COORDINATED AILERON AND RUDDER

Fig. 5-4. Typical Flight Path And Recovery Procedures For Takeoff-And-Departure Stall

changing control pressures, he will practice the stall with a *delayed* recovery. The stall is entered in the same manner as the approach-to-a-straight-ahead stall. The area is cleared and power and airspeed reduced. As lift-off speed is attained, power is increased to the recommended climb power setting. The student should observe the indications of the approaching stall, *maintain directional control*, and continue to increase the angle of attack until the stall occurs. When the stall has fully developed, pronounced buffeting will be evident and the nose will *pitch down* even though full back pressure is held. There is a tendency for the aircraft to pitch more steeply and rapidly and exhibit more rolling tendencies to the right or left in the full stall than in the approach to a stall.

The recovery is made in the same manner as with all stalls. The angle of attack is decreased by releasing back elevator pressure to lower the nose, as shown in figure 5-5. Ailerons and rudders are used in the normal coordinated manner to return the aircraft to straight and laterally level flight.

A common student error is to cause an excessive nose-down pitch attitude during recovery. At no time during the recovery should airspeeds higher than cruise airspeed be attained. Since the prime consideration in effective stall recovery is *minimum loss of altitude* consistent with positive and effective control of the aircraft, the ideal recovery should be to a laterally level attitude that will result in minimum altitude loss and not induce a secondary stall. In order to accomplish this, many instructors recommend that the airspeed not be permitted to exceed best angle-of-climb airspeed during the recovery.

While it takes several minutes to discuss the various features of this area of flight performance and the stall maneuver, the student should understand that the sequence of events, as they take place in flight, occur in a rapid sequence. For example, from the time the angle of attack is increased until recovery is made takes about 10 seconds.

DEPARTURE STALLS

Departure stalls are appropriately named since they result from inattention (or

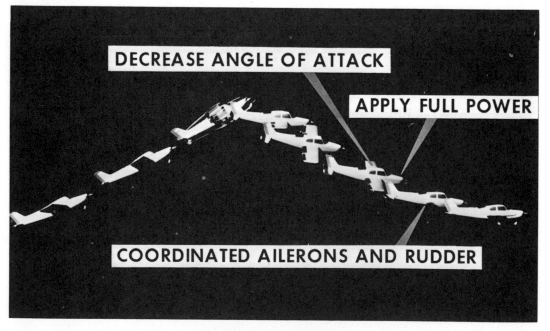

DECREASE ANGLE OF ATTACK

APPLY FULL POWER

COORDINATED AILERONS AND RUDDER

Fig. 5-5. Full-Stall Recovery

distraction) of the pilot to proper pitch and bank control of the aircraft during the departure from the airport.

The power-on, straight-ahead stall previously discussed is a basic stall. All other stalls are simply variations. Other variations will be practiced; the first of which will be with shallow-banked turns.

In the power-on departure stall, the entry to the stall is established in the same manner as for the straight-ahead stall. The difference is that as the climb is established, the student will also establish either a left or right bank. Recommended bank angles are from 15° to 20°. The student should not allow the aircraft to "roll around" within the range. He should pick one angle, 15° for example, and then strive to hold that bank constant until the stall occurs.

As the stall is approached, the angle of bank will tend to *steepen* in a left turn and become *shallower* in a right turn because of the tendency of torque and P-factor to roll the aircraft to the left. Each time the student practices stalls in each of these variations, he should strive to identify the indications of the approaching stall. The aileron and rudders should be coordinated throughout the entire maneuver.

In a power-on right turning stall, just before the stall occurs, the nose attitude will appear as shown by the visual and instrument references in figure 5-6.

The recovery from the power-on right turning departure stall is performed in the same manner as practiced earlier. The aircraft is returned to a straight-and-level attitude or slightly below while simultaneously applying all available power and using aileron and rudder pressure to level the wings. (See Fig. 5-7.)

Stall recovery is considered complete when the airplane regains straight and laterally level flight. *The takeoff and departure stalls are normally performed in a series after reasonable proficiency is obtained in performing the straight-ahead, left turning, and right turning stalls individually.*

STALLS AND FLAPS

Practicing stalls with flaps down is also a part of the flight curriculum. Usually they are practiced with half flaps and again with full flaps. Entry into the stall follows the normal procedures except that flaps are lowered to the desired setting as the airplane slows. The stall maneuver may be practiced straight ahead or in moderate banks. As the stall is approached, aileron and rudder pressure should be applied to assess the changing

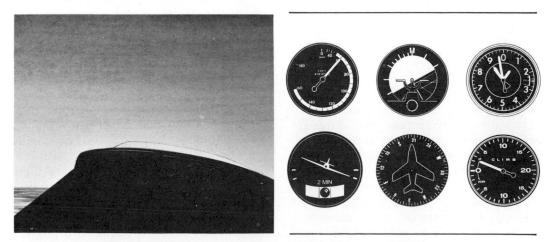

Fig. 5-6. References Just Before Right Turn Departure Stall Occurs

Fig. 5-7. References During Recovery

control responses. Between 5 and 10 miles per hour above the stall, the stall warnings will begin and continue through the actual stall until recovery has been completed.

Recovery is initiated in the approach to the partial stall or after a full stall has developed, as the instructor directs. Recovery procedures are similar to those explained for other stalls. The angle of attack should be decreased by releasing back pressure while simultaneously applying all available power and using coordinated control pressures to level the wings as the nose attitude is being set.

One *significant difference* in establishing the recovery pitch attitude when flaps are being used is that a more pronounced *nose-down* attitude is required during recovery as the amount of flaps is *increased*. Flaps will be raised only after adequate flying speed has been regained.

Stall recoveries with flaps are an *important* part of the curriculum since flap extension and retraction procedures are generally performed *close to the ground*. If flaps are retracted rapidly at slow speeds without first allowing speed to build up, it is possible to induce a secondary stall or initiate a high rate of descent. For example, if an aircraft is flying at 60 m.p.h. with full flaps and the pilot abruptly raises the flaps, the

aircraft will likely stall because the airspeed is below flaps-up stall speed of most light aircraft. However, if the pilot attains *at least* best angle-of-climb speed and then raises the flaps slowly, the chance of a secondary stall or a high rate of descent will be substantially reduced.

ACCELERATED STALLS

After the student has practiced stalls from all normally anticipated flight situations, with power on, power off, straight ahead, in turns, with flaps up, and with flaps down, he is ready to practice what is called an "accelerated stall."

The term "accelerated" has nothing to do with the rapidity with which the stall is induced. Rather, it denotes a stall which occurs at a *higher than normal airspeed* because the angle of attack is increased by an *additional load factor* normally induced during a steep turn or an abrupt increase in pitch attitude.

Accelerated stalls should not be performed in other than acrobatic airplanes at speeds *more than 1.25 times the unaccelerated stall speed*, nor with flaps extended, because of the extremely high structural loads which may result. Furthermore, abrupt pitch changes should be avoided in airplanes with extensions between the engine and propeller because of the high gyroscopic loads produced. Since the load factor on the airplane increases any time the aircraft turns or the pitch attitude is increased, all the stalls the student practiced in turns with or without power were, to a degree, accelerated stalls. However, the stall speed increase was so small that it was not easily noticed.

The increase in stall speed when performing accelerated stalls can be approximated by this formula: *stall speed increase equals the square root of the load factor*. For example, if the aircraft is placed in a 75° bank and level flight maintained, a load factor of four Gs would be imposed. The square root of four is two. Multiply the normal stall speed by two to obtain the stall speed with the increased load factor. Assuming the aircraft used in this example has a *stall speed of 63 m.p.h.*, in a 75° bank the airplane will stall at the higher airspeed of *126 m.p.h.* ($\sqrt{4}$ = 2 x 63 = 126).

Remembering the formula is not as important as the recognition that increased load factors and loads are placed on the aircraft in a steep turn or an abrupt pull-up. Normally, a maximum load factor of 1.5 Gs is considered a limit for

student practice. A 45° to 50° angle of bank will produce that load factor and will increase the stall speed approximately 1.25 times the normal stall speed; or, in other words, a 63 m.p.h. unaccelerated stall speed increases to approximately 79 m.p.h. (See Fig. 5-8.)

In preparation for an accelerated stall, the area should be *cleared* of other aircraft traffic by making clearing turns. While in the latter part of the turn, the pilot should slow the airplane to about 1½ times normal stall speed and establish a 45° angle of bank while maintaining altitude. Then, the angle of attack is progressively and briskly increased by the application of back pressure until the stall occurs. Five to 10 miles per hour above the accelerated stall speed, the stall warning indications will begin. At the time of the stall, pronounced buffeting occurs and the nose of the aircraft will appear similar to that illustrated in figure 5-9.

If the aircraft is coordinated at the stall, the nose will pitch away from the pilot as it does in level flight but in an accelerated stall, the student can expect the pitch to be gentle in most training aircraft. The recovery is initiated by releas-

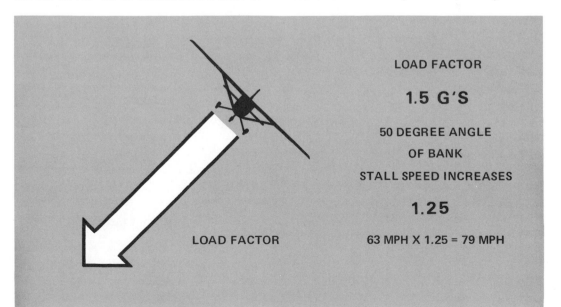

Fig. 5-8. Load Factor And Stall Speed Increase

Fig. 5-9. Nose Position In Accelerated Stall

of the turn), the *upper wing* will tend to stall first and the aircraft will *roll rapidly to the outside* of the turn.

If the aircraft is slipping to the outside of the turn at the time of the stall (the ball is to the outside of the turn), the *lower wing* tends to stall first and the aircraft will *roll to the inside of the turn;* if recovery is not initiated, the aircraft may enter a spin. In either case, the stall is "broken" by decreasing the angle of attack and using coordinated aileron and rudder pressures to roll to level flight.

ACCEPTABLE PERFORMANCE IN POWER-ON STALLS

ing back pressure which reduces the angle of attack while simultaneously adding full power. Aileron and rudder controls are used in a coordinated manner to level the wings and return to straight-and-level flight in cruise condition. (See Fig. 5-10.)

If the aircraft is slipping to the inside of the turn at the time of the stall (the ball of the turn coordinator is to the inside

As the student gains experience, his performance will be judged acceptable if correct, prompt, and smooth use of control pressures are made to achieve the desired attitudes and if he recognizes stall warnings. In approach to stalls, recovery should begin *before the nose pitches* down. In delayed recovery stalls, there should be no evidence of secondary stalls during recovery nor excessive diving. Recovery should be completed within 200 feet of the altitude at which the stall occurs.

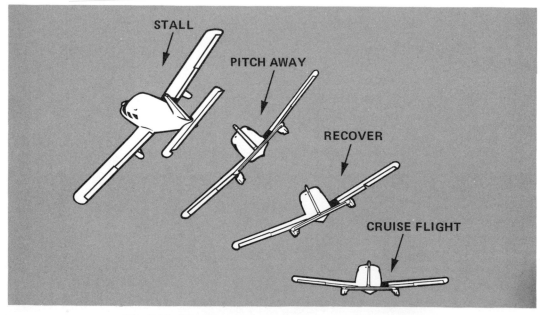

Fig. 5-10. Coordinated Recovery From Accelerated Stall

SECTION B—UNIMPROVED FIELD TAKEOFFS AND LANDINGS

The next techniques and procedures to be learned are those associated with un-improved field takeoffs and landings, commonly called short-field and soft-field takeoffs and landings.

SHORT-FIELD TAKEOFFS

During short-field practice sessions, it is always *assumed* that in addition to a short runway there is an obstruction on each end of the runway that must be cleared. The obstruction is considered to be approximately 50 feet in height. (See Fig. 5-11.)

For the short-field takeoff, the takeoff checklist is the same as used for normal takeoff procedures except that flaps are set as recommended in the owner's manual to achieve the *best angle of climb*. The recommended flap setting will vary between aircraft and may be as little as no flaps or as much as two-thirds flaps.

Repeated trials with reciprocating engine aircraft indicate that holding the brakes, applying full power, and then releasing the brakes *will not* result in a shorter

takeoff run. Therefore, the procedure recommended is to taxi along one side of the runway and make a taxiing 180° turn utilizing the *maximum* runway length and smoothly increase power through the turn, so that the aircraft completes the turn and is aligned with the runway, *full power* is being used.

Directional control is maintained with rudder pressure. The aircraft should be allowed to roll on the full weight of its wheels in an attitude that results in mini-mum drag, as shown in positions 1 and 2 of figure 5-12.

The initial takeoff roll involves *little or no use of the elevator control* beyond permitting it to assume a neutral position. Just *before* the best angle-of-climb airspeed is reached (position 3), back pressure is applied on the control wheel to smoothly and promptly attain the *best angle-of-climb attitude*. This atti-tude and airspeed should be held until the obstacle is cleared. Caution should be taken to avoid raising the nose *too soon*. A nose-high attitude will produce more drag and cause a longer takeoff roll.

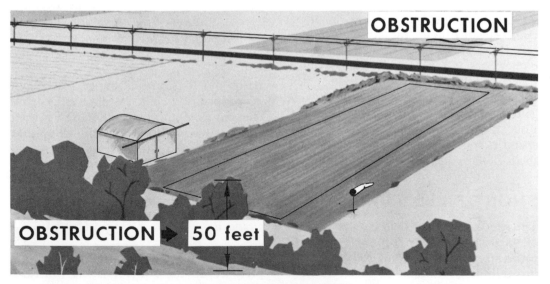

OBSTRUCTION

OBSTRUCTION ➤ 50 feet

Fig. 5-11. Typical Short Field

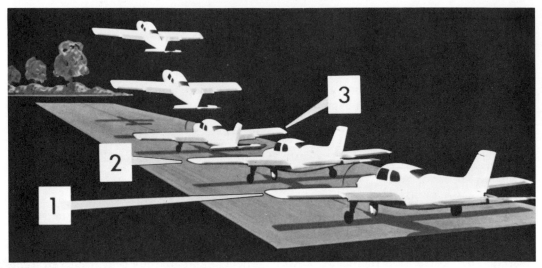

Fig. 5-12. Short-Field Takeoff Technique

After the obstacle has been cleared, the aircraft is adjusted to the normal climb attitude and power setting. If flaps are used, they should be retracted only *after* the obstacle is cleared and the *best rate-of-climb speed* has been attained. Flap retraction completes the short-field takeoff procedure and normal climbout and level-off procedures are resumed.

The question is often asked, "Why use the best angle-of-climb speed instead of the best rate-of-climb speed?" As noted in a previous section of the manual, the best angle-of-climb speed is usually slower than the best rate-of-climb speed. At the slower speed, the best angle-of-climb speed obtains the *most altitude in the shortest distance traveled*. At the higher speed, the best rate-of-climb speed obtains the most altitude in the shortest time. Since maximum altitude gain is desired within the shortest possible distance, the best angle-of-climb speed is specified for obstacle clearance, short-field takeoffs.

SHORT-FIELD LANDINGS

The successful short-field landing over an obstacle begins with *good* planning. In order to make the landing roll as short as possible, it is desirable to touch down at minimum speed and as close to the ob- stacle as possible. This means clearing the obstacle by a minimum amount at a *high angle of descent and slow airspeed.* (See Fig. 5-13.)

The early part of the approach on downwind leg and through the turn to base leg is very similar to a normal approach. During the latter portion of the downwind leg, many instructors prefer to extend one-third of the available flaps, two-thirds on base, and the remaining flaps on final approach while progressively reducing the airspeed. This enables the pilot to make the transition to short-field approach speed in smooth, easy steps. During the transition, the trim tab should be used to remove control pressure.

After the flaps are extended to the full down position and the aircraft is trimmed on final approach, the aircraft then will be in a condition very similar to that practiced in slow flight — high drag and slow speed. Therefore, it is particularly important for the student to remember the lessons learned in slow flight; *power controls altitude and elevators control airspeed.*

In the slow flight configuration, back pressure on the control wheel will raise the nose, slow the airspeed, and *increase*

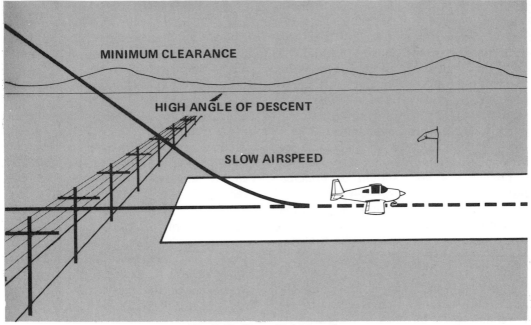

Fig. 5-13. Short-Field Landing

the rate of descent. Altitude and angle of glide are controlled by power. Power is added to *decrease* the rate of descent and angle of glide. To *steepen* the glide angle, power is reduced, and attitude and airspeed control are maintained with the elevators. On a short-field approach, minimal power may be carried throughout the final approach up to the time of roundout, or flare.

If the approach to the runway looks something like that shown in figure 5-14, the wires will be cleared. If the appearance changes to that shown in figure 5-15, power should be *added* as the air-

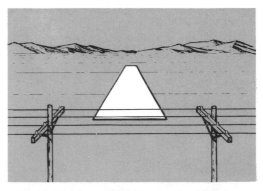

Fig. 5-15. Short-Field Approach -- Too Low

craft is *too low*; if it changes to that depicted in figure 5-16, power should be *decreased* as the aircraft is *too high*.

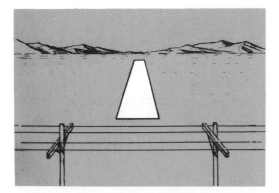

Fig. 5-14. Correct Short-Field Approach

Fig. 5-16. Short-Field Approach -- Too High

A difficult position for proper student reaction is when he is *both low and slow*. The typical reaction when low is to pull back on the wheel. However, that will only slow the speed more and *increase the rate of descent*. On the other hand, when the aircraft is low and slow, it is hard to adjust the nose position downward. It appears to the student that he is pushing himself into the obstacle or short of the field, but that is the correct response. The *nose should be lowered and power added* to increase and extend the descent. The student must think through these responses and conditions and discipline himself to make the proper attitude and power adjustments.

The pilot should plan to cross the obstacle with a speed that is equal to approximately 1.3 times the power-off stalling speed with fully extended flaps. For example, in an aircraft that stalls at 50 m.p.h. with full flaps, the speed when crossing the obstacle should be approximately 65 m.p.h. This speed is well below the normal approach speed of approximately 75 m.p.h.

After passing over the obstacle there should be no attempt to change the rate or angle of descent until flareout. The altitude at which to initiate the flareout will be approximately the same as the normal landing. If the airspeed has been properly maintained, there will be very little or no "float" after the flareout and power reduction. The airplane should touch down at *minimum controllable airspeed* in approximately the pitch attitude which results in a power-off stall. (See Fig. 5-17.)

The retraction of the flaps after touchdown must be determined by the procedure recommended in the airplane owner's manual and at the discretion and experience of the flight instructor. The aerodynamic braking produced by the flaps may be more effective than the use of wheel brakes; however, the effect of wheel brakes may be reduced materially as long as the flaps are left extended (because of the increased lift produced). Therefore, the *transition point* in the landing rollout, where flap retraction and progressive increase in brake appli-

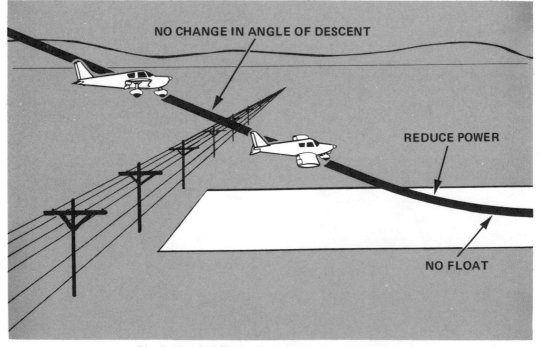

Fig. 5-17. Ideal Short-Field Approach And Landing

cation should be accomplished, *varies with the type of aircraft*. After the flaps have been retracted and as brake pressure is being increased, progressive back pressure on the control wheel may be applied to further cause the airplane to "dig in its heels" and provide more effective braking action.

The properly executed short-field landing in most light aircraft will result in an *extremely* short landing roll, and preflight planning should be exercised to insure that the aircraft can *get out of the short field* into which it has been flown.

SOFT-FIELD TAKEOFFS

The soft-field takeoff technique is used whenever the takeoff surface is covered with snow, mud, high grass, loose rocks, or the overall terrain is rough. Therefore, the objective of the soft-field takeoff is to transfer the weight of the aircraft from the *wheels to the wings* as rapidly as possible and lift the aircraft clear of the retarding effects of the surface condition.

The soft-field takeoff technique begins with the completion of the pretakeoff checklist and, as in the short-field procedures, flaps are set in accordance with the manufacturer's recommendations found in the owner's manual. After completing the checklist, the aircraft should be briskly taxied toward the takeoff position to prevent the aircraft from "bog-

ging down" while maintaining *full* elevator back pressure to place *minimum* weight on the nosewheel.

At no time during the taxi onto a soft runway surface should the aircraft be allowed to stop; rather, alignment with the center of the takeoff area should be performed while taxiing and the throttle advanced as rapidly as the engine will accept power without faltering. As speed picks up, some of the elevator back pressure must be relaxed; however, sufficient pressure should be maintained to lift the nose gear clear of the surface as soon as possible. When this is accomplished, the aircraft will no longer benefit from the effects of nosewheel steering, and prompt positive rudder control pressure is necessary in order to maintain directional control (as in slow flight at minimum controllable airspeed).

The nose position is considerably higher than in a normal takeoff and may require the pilot to sight along the edge of the cowling in order to have adequate forward vision. The high angle of attack and *ground effect* (ground cushion) will result in a lift-off speed that is actually *lower* than the normal stall speed in the same configuration. (See Fig. 5-18.)

On a rough surface, it is possible for the aircraft to skip or bounce into the air before the full weight of the airplane can be supported aerodynamically. Therefore, it is important for the pilot to hold

Fig. 5-18. Soft-Field Takeoff

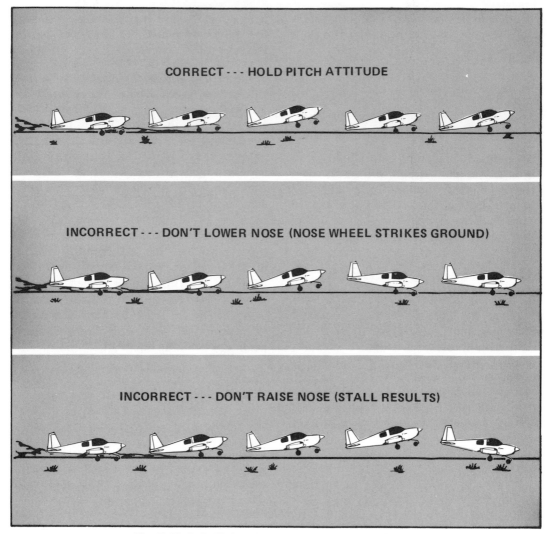

CORRECT - - - HOLD PITCH ATTITUDE

INCORRECT - - - DON'T LOWER NOSE (NOSE WHEEL STRIKES GROUND)

INCORRECT - - - DON'T RAISE NOSE (STALL RESULTS)

Fig. 5-19. Soft-Field Takeoff --- Pitch Attitude Control

the pitch attitude as constant as possible (an important application of slow flight at minimum controllable airspeed). As shown in the top portion of figure 5-19, if the aircraft settles back to the surface, it can continue its acceleration to take-off speed. Permitting the nose to lower after a bounce may cause the nosewheel to strike the ground with resulting damage, as shown in the middle portion of figure 5-19. On the other hand, sharply increasing the pitch attitude after a bounce may cause the aircraft to stall, as shown in the bottom portion of the illustration.

Once the aircraft is airborne at the minimum airspeed, it should be held *just* *clear of the surface* and the *angle of* *attack slowly reduced* in order to *accel-* *erate* to the best angle-of-climb airspeed *before further climb* is initiated. As best angle-of-climb airspeed is obtained, a gradual climb may be initiated and the aircraft permitted to accelerate to best rate-of-climb airspeed. At this time, the flaps may be slowly retracted (if they were used to perform the soft-field take-off).

SOFT-FIELD LANDINGS

The soft-field landing also assumes that the runway surface is covered with snow, mud, high grass, loose rocks, or that the overall terrain is quite rough; therefore,

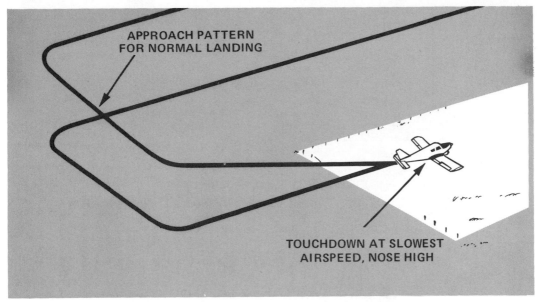

APPROACH PATTERN
FOR NORMAL LANDING

TOUCHDOWN AT SLOWEST
AIRSPEED, NOSE HIGH

Fig. 5-20. Soft-Field Landing

the objective is to support the weight of the aircraft during the landing roll *with the wings as long as possible* and delay the weight transfer to the wheels during the rollout until the aircraft attains the slowest possible speed. The airspeed used for the *short-field landing* is equally appropriate for the soft-field landing, and a normal landing approach is recommended. There is no reason for a steep approach path *unless obstacles* are present in the final approach course. (See Fig. 5-20.)

A small amount of power should be maintained throughout the flareout to assist in decreasing the rate of descent while permitting the speed of the aircraft to be reduced to the lowest possible value. Use of power will also permit the aircraft to contact the ground as softly and smoothly as possible.

The use of flaps should be according to the recommendations in the owner's manual, but governed also by the existing conditions. If there is an obstacle to clear and a short runway, flaps *should be used as recommended for a short-field landing.*

On soft surfaces, deceleration is rather rapid after touchdown; braking is not usually required and may even be undesirable or lead to a nose-over tendency. In order to prevent nose gear damage or prop damage, it is desirable to hold the nosewheel off the ground as long as possible by progressively increasing the elevator back pressure as the aircraft is decelerated during the landing rollout. This procedure is illustrated in figure 5-21.

COMBINING TAKEOFF AND LANDING TECHNIQUES

Throughout the discussion, one specialized takeoff or landing technique has been dealt with at a time for the sake of simplicity and ease of description. However, it is possible to have situations where it is necessary or desirable to use a combination of two or more techniques. For example, the pilot could encounter a situation involving a *soft field* that has *obstacles* at the end of the runway and a *crosswind component.* A combination of soft- and short-field procedures and crosswind techniques would be required.

All possible situations cannot be covered in one manual. There will be times when the techniques learned must be modified to fit existing conditions. The flight instructor will provide guidance for addi-

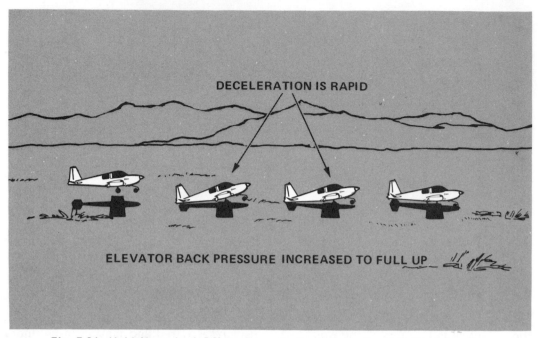

Fig. 5-21. Hold Nosewheel Off As Long As Possible During Soft-Field Landing

tional techniques the student may be required to use, and as the pilot gains experience, he will develop proficiency in *combining* the various techniques to fit the existing flight conditions.

For example, modification of landing technique is required when the winds are high, gusty, and variable. In this situation, the final approach airspeed is higher and the touchdown attitude is less nose high. Also, the touchdown should be at a higher airspeed to provide positive and prompt control responses throughout the flareout. Similarly, in high gusty winds, takeoff airspeeds should be higher and the nosewheel held on the ground longer to enhance directional control.

The various types of takeoffs and landings are taught so that the pilot may be able to utilize the full performance capability built into his aircraft and control the aircraft throughout its performance envelope with security and confidence. When he has achieved this mastery of his aircraft, he will be able to approach a wide variety of existing conditions with the cool confidence of a professional.

ACCEPTABLE PERFORMANCE FOR UNIMPROVED FIELD TAKEOFFS AND LANDINGS

Performance will be evaluated on the basis of planning, smoothness, directional control, and accuracy. The student should recognize and maintain the proper attitude and airspeed required for each phase of the takeoff and landing. Airspeed should be maintained within five knots of that required. Extreme caution and positive control must be exercised during flap retraction on roll-out.

SECTION C—EMERGENCY LANDING PROCEDURES

Modern aircraft engines are extremely reliable and actual mechanical malfunctions are a rare occurrence. However, a situation where the pilot is flying with "brain in neutral" and runs out of gas, or the more remote possibility of an actual engine component malfunction, may require the pilot to make an off-airport emergency landing or forced landing. Five general steps used to cope with such a situation are listed in sequence:

1. *Maintain altitude* and set up the best glide speed.
2. *Scan the immediate area* for a suitable field.
3. *Turn to a heading* that will take the aircraft to that field.
4. *Attempt to determine the cause* of the power failure and, if possible, restart the engine.
5. *Set up a landing approach* to the selected field.

MAINTAIN ALTITUDE AND SET UP GLIDE SPEED

If the "fan should stop," the pilot should attempt to *maintain his altitude* by the application of back pressure on the control wheel as the aircraft is slowed to the best glide speed. This speed is often referred to as the "best glide speed" and is normally specified in the aircraft owner's manual. This speed also is *usually close to the approach speed* without flaps, which is listed for most light training aircraft.

As the best glide airspeed is attained, the aircraft should be trimmed since this procedure will relieve the pilot from holding control pressures to maintain the proper attitude and airspeed. If the airspeed should be *below* the best glide speed at the time of power failure, the nose should be *immediately lowered* to the best glide attitude and the aircraft retrimmed. If the flaps are also down, the airspeed should be increased to a safe flap retraction speed (best

glide speed, if possible), the flaps retracted, the proper attitude established, and the aircraft retrimmed.

If engine operating procedures require carburetor heat during a prolonged glide, the instructor will normally apply carburetor heat as he reduces the power in simulated forced landing practice. If he does not "pull on" carburetor heat and heat is required, the student should apply heat immediately as the power is reduced.

SCAN THE AREA FOR A SUITABLE FIELD

The selection of the field should be made *within the immediate area*. A rule of thumb to follow is: *do not select a field that is at more than a 45° angle from a vertical line beneath the aircraft.* In much simpler terms, select a field that can be easily evaluated for obstructions, terrain, etc. and one that can be "made" from the present altitude. The pilot who selects a "beautiful" field 10 or 15 miles from his present position and arrives over that field to find it covered with rocks, trees, and criss-crossed with several ditches or fences would have been better off to have selected a field within the immediate area over which he was flying.

Experience and familiarity with the general terrain in the training area will help in the selection of the field. There are many variables to consider, including *wind direction and speed, length of field, obstructions, and the surface condition of the field.* These various factors must be evaluated and a decision made that provides the greatest possibility of a successful outcome.

Ideally, a long field lying into the wind with a firm, smooth surface and free of obstructions would be the most desirable; however, all of these features are seldom available and some undesirable characteristics must be accepted. On one occasion, it may be better to accept a crosswind landing on a long field rather than attempt to land on a very short field lying into the wind. Similarly, on another occasion, a downwind landing with light winds and no obstructions may be preferable to a landing into the wind with numerous obstructions ahead.

Each case must be judged on its own merits, and the flight instructor will provide guidance for the student regarding special terrain and seasonal effects in the training area. He will also critique the student's choice of a field after each practice approach.

HEAD FOR THE FIELD

If the previous suggestions are followed, the selected field should be within easy gliding distance. This means that the aircraft should be headed directly for the field and any excess altitude should be dissipated in a *spiralling approach over the field.* From that vantage point, the pilot will be in a good position to carefully observe the field for wires, fences, holes, tree stumps, ditches, or other hazards that may not be easily observed from a greater distance.

It is *inadvisable* to make *S-turns* or to circle away from the field, and then try to make one long straight-in glide to the field. The estimation of glide distance from a faraway point is difficult even for experienced pilots. More often than not, these attempts end in an undershoot or overshoot because of misjudgment. Moreover, if the student elects a long straight-in glide, he is *committed.* If he discovers he will be short, he has no method to adjust the glide. A circling approach over the field improves the ability to plan and make adjustments for altitude and constantly keeps the pilot in a position from which he can "make the field."

Throughout the glide, the instructor will keep the engine *cleared.* In order to do

this, he will make short applications of power to warm the engine periodically and prevent the engine from "loading up."

ATTEMPT TO DETERMINE CAUSE OF POWER FAILURE AND RESTART ENGINE

There is no hard and fast rule as to *when* this particular step should be accomplished. However, it should be performed *as soon* as the glide is set up, the field selected, and the aircraft headed for that field. As mentioned previously, complete and immediate power failures generally result from some pilot error, such as running out of fuel, and it can be extremely embarrassing for a pilot to make a forced landing only to find that he could have continued flight had he switched the fuel selector from the empty tank to one that contained fuel.

Usually, the sudden engine power loss is caused by a fuel problem; therefore, the pilot should check to make sure the *fuel selector* is ON and set to a tank that *has fuel.* Most instructors require the student to switch (or simulate switching) tanks automatically as part of the procedure.

Next, the *mixture* should be checked to see that it is in the RICH position. *Carburetor heat* also should be applied early in the procedure to determine if ice has formed and attempt to remove it. Then the *magneto switch* should be checked to see that it is in the BOTH position. If none of these checks reveals the cause of the power failure, the pilot should scan other instruments and make a general inspection inside the cabin to obtain some clue as to the cause of the failure. If the cause of the power failure *is discovered and remedied*, engine restart *does not require the use of the starter* since the propeller continues to turn or "windmill" in a power-off glide.

The student will be encouraged to be methodical, perform his checks in a definite sequence, and to take time to be thorough with the cockpit check. It is a bit startling when engine power is reduced suddenly, but from gliding practice, the student should realize that the aircraft is under perfect control at glide speeds. It may seem as though it takes minutes to go through a cockpit check, but actual tests reveal that it requires only about 15 seconds to make an *unhurried check*. Therefore, if the rate of descent is 600 feet per minute, the most altitude that should be lost while performing the check is about 150 feet.

SET UP A LANDING APPROACH TO THE SELECTED FIELD

360° OVERHEAD APPROACH

If the student has been spiralling downward over the field, he should attempt to complete the last spiral *at approximately 1,500 feet, headed into the wind, and directly over the intended landing spot*. This type of spiral approach is shown in figure 5-22. From this position, the student can continue the circling descent in order to place himself at the familiar downwind leg position used for normal landings.

THE 180° SIDE APPROACH

Ideal planning should place the aircraft at the 180° point at the normal traffic pattern altitude above the ground. (See Fig. 5-23.) From that point, the approach is like a normal power-off approach at the home airport albeit with an added "perspiration factor." The student should be able to use all of the normal clues to turn base leg, judge position at the *key point*, and turn to final.

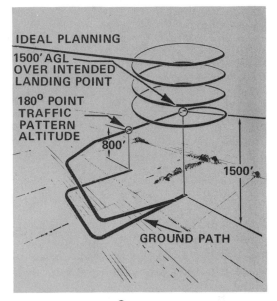

Fig. 5-22. 360° Overhead Approach

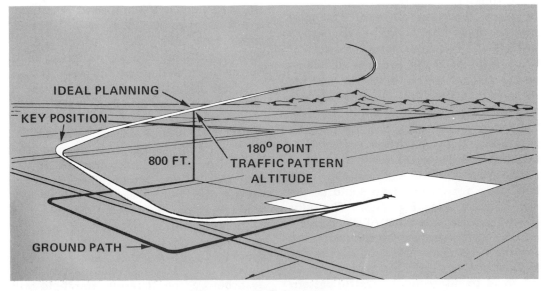

Fig. 5-23. 180° Side Approach

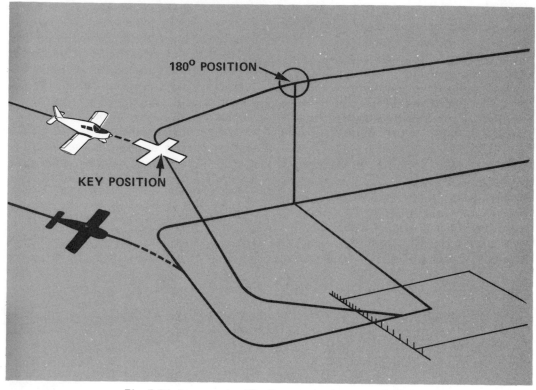

Fig. 5-24. Intercepting The Pattern At The Key Position

The value of this procedure is that it places the pilot in a position where he has familiar points from which to make judgments on glide angle, speed, and distances.

Unfortunately, it is not always possible to reach the ideal position. The pilot may have to use a right-hand pattern instead of a left-hand pattern because of his position. If the altitude at which power failure occurs is too low or the distance too great, the student may not be able to arrive at either a left or a right 180° position. If this should happen, the alternative is to make the approach so that the aircraft can *intercept* the normal traffic pattern. For example, the next best place for interception could be the key position. (See Fig. 5-24.) The pilot should use the technique of visualizing a normal traffic pattern overlayed on the chosen field. Then he should consider the altitude and his position and plan to fly so as to intercept the traffic pattern at the earliest point.

During the approach, flaps should be used as required. It is recommended that full flaps be used only *after turning onto final approach*, and when the pilot has the assurance that he will reach the intended field.

During emergency landings, it is imperative that the pilot remember the distance traveled *in the flare* during landing. While discussing the aiming point used during normal landings, it was pointed out that the glide angle aiming point is *short of the actual touchdown point*.

Figure 5-25 shows a ditch with the desired landing point just beyond the ditch. The aiming point must be on the near side of the ditch. When properly executed, it will appear that the aircraft will touch down short of the intended point and, unless the student makes a conscious effort, it is difficult to aim short of the obstruction. (See Fig. 5-26.)

Just before the flareout, the field will appear like that shown in figure 5-27.

Fig. 5-25. Typical Final Approach View of Emergency Landing Approach

Fig. 5-26. Aiming Point and Probable Touchdown Point

During the flare, the aircraft will glide across the ditch and, as it slows, will land at the desired spot, as shown in figure 5-28. A profile view of the final portion of the approach just described would appear as illustrated in figure 5-29.

Fig. 5-27. View Just Before Flareout

Fig. 5-28. View At Touchdown

Fig. 5-29. Profile View of Final Approach

PROGRESSION OF DIFFICULTY LEVEL

The forced landings given prior to solo by most instructors consist of only those types that require a straight glide or a turn of not more than 90°. When the student is confronted with such a problem, the selection of the proper field and approach factors are within his capability and experience levels. Immediately after solo, the 180° side approach may be added to the emergency landing problem and, as flight training progresses, forced landings can be expected from higher altitudes where a spiralling descent and a 360° overhead approach may be included.

THE GO-AROUND

If the selected field is an approved landing area, the instructor may require the student to proceed and land. If the field is not approved, a go-around should be initiated no lower than 200 feet above ground level (AGL). From the altitude, it should be apparent whether the out-come of the simulated emergency landing would be successful or not. No attempt should be made to go close to obstructions nor should flight be continued closer to the ground.

It should be clearly understood between the instructor and the student as to *who will perform the go-around*. If the instructor announces, "I've got it," the student should relinquish complete control of the aircraft. Once the instructor has assumed control, the student should not touch the flight controls, retract flaps, or perform any other function unless *directed* to do so.

To initiate the go-around, all available power should be applied. The carburetor heat control should be placed in the COLD position. The airspeed should be at or above the best angle-of-climb speed before flap retraction is initiated. When well clear of all obstructions, a normal climb can be resumed and, in order to conform with normal traffic pattern procedures, no turns are recommended below 400 feet AGL.

ACCEPTABLE PEFORMANCE DURING FORCED LANDING PROCEDURES

Acceptable performance is characterized by the prompt recognition of the power loss and the ability to establish the attitude which will result in the best glide speed. Procedures should be followed methodically and precisely, and the normal traffic pattern intercepted and adhered to as soon as practical. The airspeed should be held within 10 miles per hour of the recommended approach speed. At the 200-foot AGL point, when the go-around is initiated, it should be evident that the aircraft could safely land in the selected field.

EMERGENCY LANDING PHILOSOPHY

Since most practice emergency landing approaches terminate in a go-around, it is possible for the student, and even the instructor, to fall into the habit of considering the procedure as just another training exercise. Certainly, getting "up tight" is not recommended; however, a serious attitude and *the assumption that each simulated emergency landing will actually result in a landing should be adhered to*. Careful cultivation of this assumption will prepare the student for the remote but ever-present possibility that he may actually perform an emergency landing during his piloting career.

OTHER EMERGENCY OPERATIONS

In addition to emergency landing practice, the student must have the required knowledge to cope with certain other situations. During normal flight operations, the instructor will simulate or state a given problem and then help the student take prompt, corrective actions. These problems will include: partial loss of power, engine roughness, carburetor ice, and fuel starvation. Also, systems malfunctions on such items as landing gear, wing flaps, trim tabs, and an open door in flight will be covered.

CHAPTER

GROUND REFERENCE MANEUVERS

6

INTRODUCTION

The maneuvers practiced early in the flight training program are done at altitudes *well above* traffic pattern altitudes. These maneuvers are practiced primarily to learn how the aircraft feels and responds in various configurations and flight conditions and to develop the student's ability to control the aircraft with smoothness and precision. Once the student has gained elementary control of the aircraft, he will be introduced to problems that require him to *divide his attention between flying the aircraft and following prescribed paths over the ground.* The diversion of attention to objects on the ground introduces the student to flying the aircraft by *subconscious* responses and helps establish these responses through repetition. Since diversion of attention is developed by these exercises, they are of primary importance in preparing the student pilot for maneuvering in the traffic pattern, and making approaches, landings, and departures where a division of attention

within and outside of the aircraft is required for the safe conduct of flight.

Ground reference maneuvers, as they are called, are performed at approximately 600 feet above ground level (AGL) which is low enough so the student can clearly see the effects of *wind and heading changes* on his path over the ground. Simpler ground reference maneuvers are taught early in the program, and as the student gains experience, more complex maneuvers will place additional demands on his planning, timing, coordination, and subconscious control of the aircraft.

EXPLANATION OF TERMS

Frequently the terms *heading, crab,* and *track* will be used in this chapter. In order to clear up any question about terms that may be used, references to "heading" pertain to the aircraft's magnetic compass heading. On the other hand, "track" pertains to the *path* that the aircraft actually makes *over the ground.* "Crab" means turning into the wind to *correct for drift* and make good the intended track.

SECTION A – TRACKING A STRAIGHT LINE BETWEEN TWO POINTS AND CONSTANT RADIUS TURNS

A basic fundamental requirement to perform any ground reference maneuver is the capability of flying a straight line over the ground between two points. (See Fig. 6-1.) It will be observed that with *no wind*, if the aircraft passes over a position on the ground (point A) and is pointed at a second point (B), it will *fly directly to and pass over* the second point. However, when a crosswind is present, if the aircraft is pointed at the second point when over the first point on the ground and a constant heading held, the aircraft will *drift to one* side and miss the second point by a considerable margin. How far to the side the aircraft drifts will depend on the *strength of the wind* and how directly *perpendicular* the crosswind is to the flight path. (See Fig. 6-2.)

WIND DRIFT CORRECTION TECHNIQUES

If the pilot simply adjusts the aircraft heading periodically in order to aim at point B, the flight path over the ground, or track, will assume a curved shape until the aircraft ends up headed into the wind, as shown in figure 6-3.

The *accepted method* used to correct for the effects of wind is to make a coordinated turn *into the wind* so that the nose of the aircraft no longer points toward the intended point on the ground, but rather is inclined in the direction *from which* the wind is blowing. The magnitude of this coordinated turn is determined by wind speed and direction. *Crabbing*, as this maneuver is called, gives the appearance that the airplane is flying sideways; however, this is an illusion since the airplane is still flying *directly into* the oncoming airstream (relative wind) just as in a calm wind situation. (See Fig. 6-4.)

The beginning pilot may recall that when solving cross-country navigation problems, he computes the course between any two points, then obtains a forecast

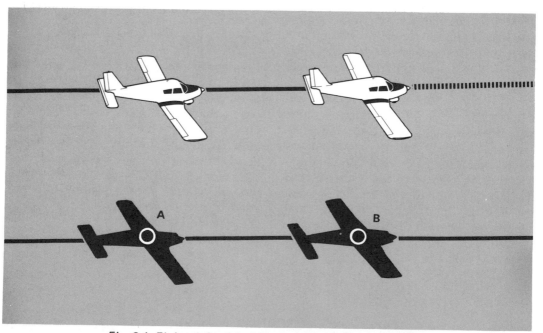

Fig. 6-1. Flying A Straight Line Over The Ground — No Wind

Fig. 6-2. Effect Of Wind On Aircraft Ground Track

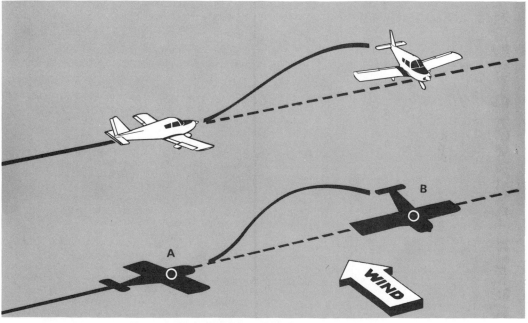

Fig. 6-3. Incorrect Wind Drift Correction Technique

of wind direction and speed. These known quantities, together with a calculated true airspeed, can be set on a flight computer and a specific heading and wind correction angle computed.

However, for ground reference maneuvers, the amount of wind correction is determined by the trial-and-error method, as shown in figure 6-5. That is, the aircraft is established on a heading and the drift observed, as shown between positions 1 and 2. Then, as illustrated at position 2, the pilot should turn the aircraft to compensate for drift,

Fig. 6-4. Crab Correction For Wind Drift

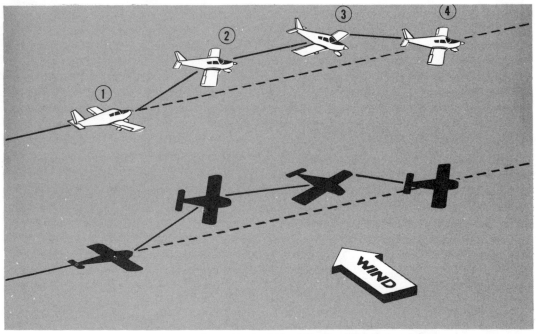

Fig. 6-5. Trial-and-Error Method Used To Establish Correct Crab

hold that new heading, and observe the new ground track, as diagrammed between positions 2 and 3. If the aircraft is still drifting from the desired track, the pilot should make another heading correction, as seen at position 3, hold this heading, observe the track, and continue this process through a series of small cor-

rections until he has arrived at the heading necessary to maintain the desired track.

COORDINATION

It should be emphasized that all turns made while performing ground reference maneuvers are *coordinated* turns. Aileron, rudder, and elevator pressures are applied as in earlier high-altitude practice. There should be no tendency to *skid* the nose with the rudder in order to change the heading slightly, nor should the pilot attempt to hold a crab angle with *rudder pressure.*

DETERMINING DRIFT

To determine the *amount* of drift, a *point on the ground* is aligned with a reference sighting point in line with the pilot's eye. As explained earlier, this imaginary line of sight passes from the pilot's eye, *through the windshield* (at a given distance above the aircraft cowling), and intercepts the horizon. If there is no drift, the point on the ground and the reference point (as it penetrates the windshield) will stay in alignment as the

aircraft proceeds toward the point on the ground. (See Fig. 6-6.) On the other hand, if the aircraft is drifting, the reference line will *drift downwind* from the point on the ground, as illustrated in figure 6-7.

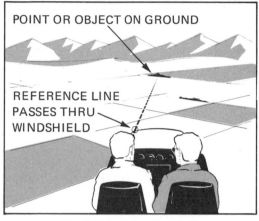

Fig. 6-6. Sighting For Drift

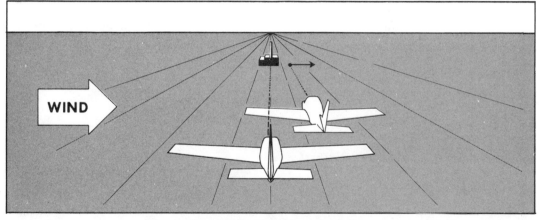

Fig. 6-7. Visual Evidence Of Drift

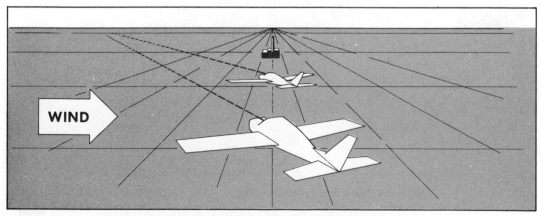

Fig. 6-8. Wind Drift Properly Corrected With Crab

To establish the proper drift correction angle, the reference line of sight must be moved to the *windward* (downwind) side of the point on the ground over which the intended track must pass. As the aircraft continues toward the ground point, the drift correction angle (crab) tends to give the appearance that the aircraft is sliding along sideways; and, indeed, this is true in *reference to the ground.* When the student has applied the right amount of drift correction, the angle *between* his reference line of sight and the point over which he intends to pass will remain fixed, as shown in figure 6-8.

The ability to hold a constant heading is important in maintaining a successful track. Therefore, if the pilot's normal variations in heading control are excessive, he will not be able to maintain his desired course over the ground.

The student can prepare for flying ground reference maneuvers by *deter-mining* the direction of the wind before he leaves the airport and *verifying* the direction and velocity as he approaches the practice area. The runway in use, windsock, smoke, and wind streaks on the water are valuable clues. Also, the wind produces advancing wave patterns on grain fields and leans rows of trees away from the wind. All of these clues, then, are helpful in determining wind direction and are illustrated in figure 6-9.

By the time a road or an imaginary line between two points is selected for practice tracking a straight line between two points, the student should have the direction and speed of the wind in mind so he can make an initial estimate of the crab angle required. *Coordinated* turns are used to make small corrections to the initial crab angle. The student should avoid making small heading changes by the application of rudder pressure only (he will also avoid his instructor's "wrath"). The track over the ground will

= DIRECTION FROM WHICH WIND IS BLOWING

Fig. 6-9. Useful Clues To Wind Direction And Velocity

reveal if the student has the proper drift correction angle established.

While performing these basic tracking problems, the student will also gain some "feel" for his speed over the ground. The time it takes to go between any two points depends on *groundspeed* rather than true airspeed. As explained earlier, if there is a tailwind, the groundspeed is faster than true airspeed; with a headwind, the groundspeed will be slower than the true airspeed. The ability to recognize groundspeed *changes* is important because it helps the pilot anticipate and plan his flight path in order to make precise ground reference maneuvers.

Before proceeding further in this discussion, let it be stated that the explanation and guidelines given can be very helpful in understanding the theory of

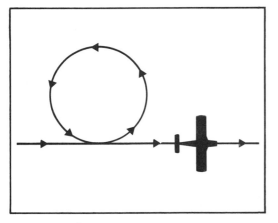

Fig. 6-10. Ground Track With Constant Bank (No Wind)

drift correction necessary to fly precise ground reference maneuvers. However, once the student actually *observes* these effects in the airplane the "whole picture" seems to become much simpler than the explanation. This is somewhat akin to describing how to perform a basketball "lay up shot" in a step-by-step fashion to someone who has never seen nor performed this shot as opposed to demonstrating the shot and then requiring the individual to perform it.

CONSTANT-RADIUS TURNS

The effects of wind on a *turning* aircraft are not quite so obvious as in straight flight. Figure 6-10 illustrates that if an aircraft turns at a constant angle of bank in a *calm wind condition*, the track over the ground will be circular and the aircraft will end up over the point where the maneuver began.

However, a turn performed at a constant angle of bank in a headwind will produce an elongated elliptical ground track, and the turn will be completed on the original track, but *behind* the point where it started. This distorted ground track is shown on the left in figure 6-11. Similarly, a constant-angle-of-bank turn performed in a direct crosswind will produce a distorted ground track and be completed *abreast* of where the turn started, but to the downwind side, as illustrated in the right portion of figure 6-11.

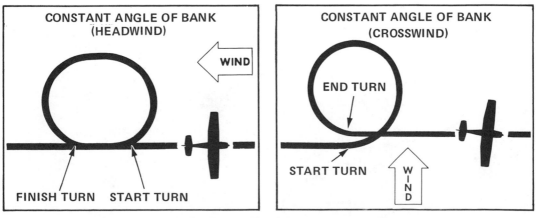

Fig. 6-11. Effect On Ground Track Of Headwind And Crosswind

If the pilot desires to make a turn in a wind and keep a *constant radius* or distance from a point, *the angle of bank must be varied to change the rate of turn.*

Most flight instructors demonstrate the effects of wind on turns performed at a constant angle of bank at an altitude of approximately 600 feet above the ground. At this altitude, the effects of the wind and resulting distortion of the ground track are readily apparent.

The *steepest angle of bank in a constant-radius turn* is at the point of *greatest groundspeed.* The steeper bank and higher rate of turn are required to turn the aircraft faster since its speed over the ground has increased. This point of greatest groundspeed is where the aircraft is flying *directly downwind* and the true airspeed *plus* the wind speed results in the higher groundspeed. These relationships are shown in figure 6-12.

In contrast, the *shallowest angle of bank* is at the point of *slowest groundspeed.* The shallower bank and slower rate of turn are necessary because the aircraft is passing over the ground more slowly. As

shown in figure 6-13, this point of slowest groundspeed occurs when the aircraft is going *directly into the wind.* At this point, the true airspeed *minus* the wind speed results in the slower groundspeed.

It is sometimes *mistakenly* thought that the steepest and shallowest banks occur when the aircraft is flying directly crosswind. However, a careful inspection of the turn reveals that the aircraft's groundspeed is the *same* since the crosswind adds neither a headwind nor a tailwind component to affect the groundspeed. In other words, the groundspeed is an average — slower than when flying directly downwind and faster than when going into the wind. Therefore, the angle of bank is a value *midway* between the steep bank required downwind and the shallow upwind bank. (See Fig. 6-14.)

To make a turn that traces a perfect circle over the ground, the angle of bank must be *constantly changed* throughout the entire circular maneuver, as seen in the top portion of figure 6-15. The bank is gradually steepened from its shallowest point until the steepest bank is achieved when the aircraft is flying directly downwind. From the point

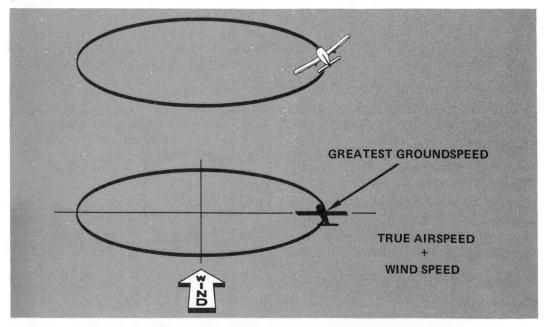

GREATEST GROUNDSPEED

TRUE AIRSPEED
+
WIND SPEED

WIND

Fig. 6-12. Steepest Angle Of Bank Is Where Groundspeed Is Greatest

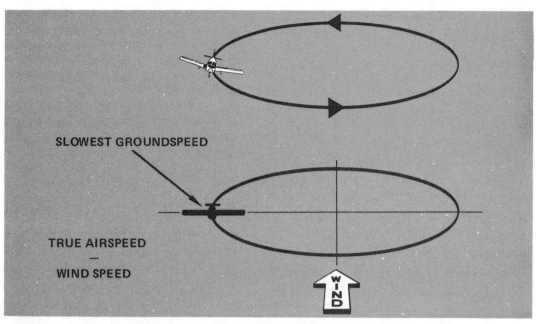

Fig. 6-13. Shallowest Angle Of Bank Is At Point Of Slowest Groundspeed

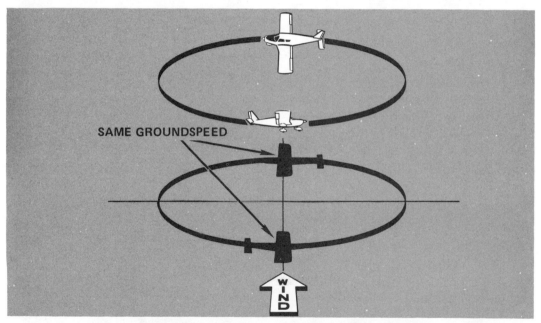

Fig. 6-14. Same Angle Of Bank Required On Upwind And Downwind Sides Of Turn

where the aircraft is flying directly downwind, the bank is gradually shallowed.

It should also be pointed out that with a constant-radius turn in a wind, the aircraft must be flown with a *drift correction angle*, or crab. As illustrated in the bottom portion of figure 6-15, on the *downwind side* of the turn, the nose must be *crabbed toward the center* of the circle, and on the *upwind side*, it must be crabbed away from the circle's center.

As the pilot looks out his side window on the upwind side of the circle, the center of the circle appears *behind* the lateral axis of the airplane, usually located near the front spar of the wing. (See Fig. 6-16, top portion.) In contrast, as the pilot looks out the side window on the downwind side, the center of the circle appears to be *ahead* of the lateral axis as shown in the center illustration. These appearances are most pronounced when directly crosswind. When flying directly upwind or downwind, the lateral axis will appear to be nearly in alignment with the center of the circle. (See Fig. 6-16, bottom portion.)

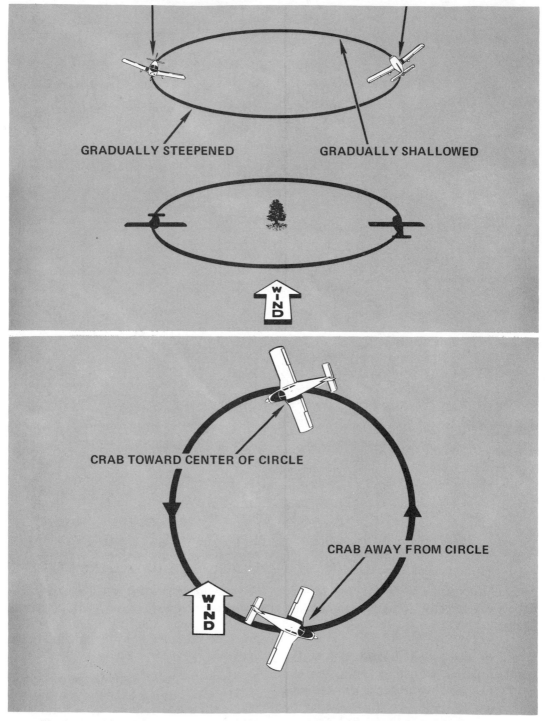

Fig. 6-15. Bank Angle And Crab Corrections Required For Constant-Radius Turn In Wind

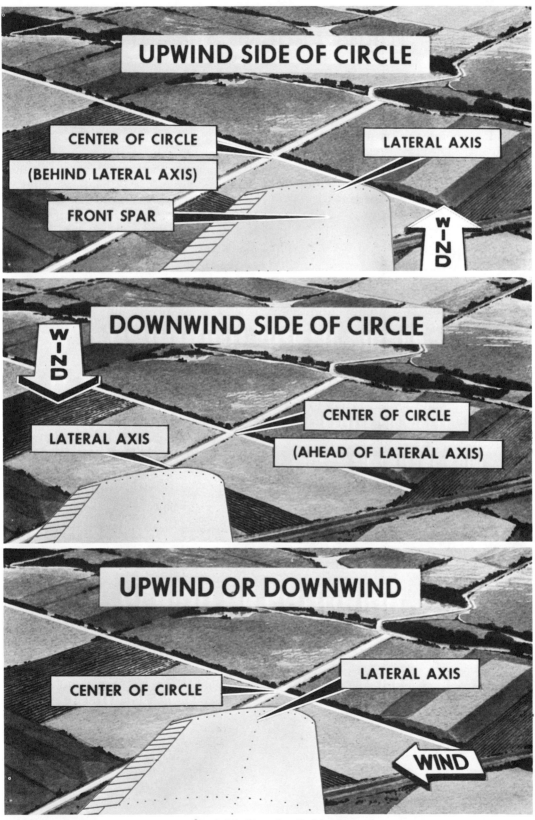

Fig. 6-16. Changing References

SECTION B—S-TURNS ACROSS A ROAD AND TURNS AROUND A POINT

Two of the most commonly practiced ground reference maneuvers are S-turns across a road and turns around a point. It should be emphasized that the turns around a point *are* ground reference maneuvers. The steep turns practiced earlier at altitude were made at a constant angle of bank with no attempt to describe a particular ground track.

S-TURNS

The first step in performing S-turns across a road is to select a road that is *perpendicular* to the wind direction. The objective is to fly two perfect half circles of equal size on opposite sides of the road, as shown in figure 6-17. As with other ground reference maneuvers, the practice altitude is approximately 600 feet above ground level. Further examination of the illustration reveals that the S-turn *really* is two halves of a constant-radius turn in which the direction is changed midway through the maneuver.

Figure 6-18 represents the aircraft and the resulting ground track as they would appear from above. As with most ground reference maneuvers, the pilot should enter the maneuver on a downwind heading. As the road is crossed, he should immediately roll into a bank. Since the aircraft is flying downwind, it is at its *highest groundspeed;* therefore, the angle of bank selected at this point will be the *steepest* used throughout the maneuver. As shown in the left portion of the figure, the bank is then gradually reduced to trace a half circle. The rollout should be timed so the aircraft is rolled level, headed directly upwind, and is perpendicular to the road just as the road is crossed.

As the road is crossed, the pilot should smoothly and gradually roll into a turn in the opposite direction. This portion of the maneuver requires the *shallowest bank* since the aircraft is headed upwind and has the *slowest groundspeed.* The bank is gradually steepened throughout

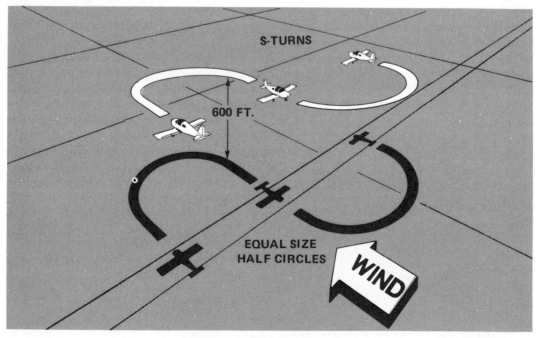

S-TURNS

600 FT.

EQUAL SIZE
HALF CIRCLES

WIND

Fig. 6-17. S-Turns Across A Road

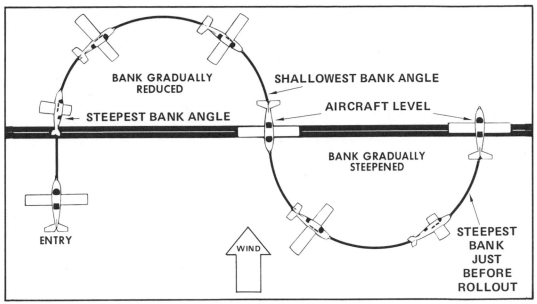

Fig. 6-18. Maintaining The Track By Varying The Bank Angle

the turn to continue tracing a half circle track in the same shape and size as on the opposite side of the road. The steepest bank is attained and the "S" completed as the aircraft crosses the road. The pilot should level the wings as he crosses the road perpendicularly, be headed directly downwind, and be ready to repeat the "S", if desired.

Figure 6-19 illustrates that, as in constant-radius turns, the aircraft must be "crabbed" to maintain the desired ground track. During the turn from the downwind portion of the "S", the nose of the aircraft must be crabbed to the *inside* of the turn. Similarly, during the upwind turn, the aircraft must be crabbed to the *outside* of the turn to maintain proper drift correction.

TURNS AROUND A POINT

The technique for performing turns *around a point* should not be confused with the steep turns performed at altitude with a *constant angle of bank*. Rather, turns around a point are ground reference maneuvers performed at an altitude of approximately 600 feet AGL. They are an application of the constant-radius turn in that the pilot must per-

form turns around a preselected point on the ground.

In addition to maintaining a constant radius from the point, further objectives are to maintain a constant altitude, maintain orientation, and roll out on the initial heading after two full turns. The maneuver and resulting ground track is represented diagrammatically in figure 6-20.

This maneuver is first introduced using a medium bank at the altitude specified, as higher altitudes make the planning of the flight track about the point more difficult. The point selected as a reference should be prominent, easily distinguished, and yet small enough to establish a definite location.

Pointed trees, isolated haystacks, or other small landmarks are sometimes used, but are not as effective as the intersection of two roads or fencelines. Roads or fencelines prove much more desirable because during *right* turns, the wing may momentarily block the pilot's view of the reference point. However, if the pilot has selected a road or fenceline intersection, he can mentally project these lines to their logical intersection and thereby maintain his orientation. In addition, the point selected should be in

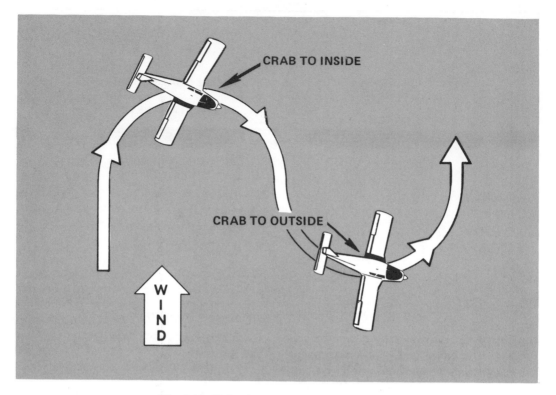

Fig. 6-19. Maintaining The Track With A Crab

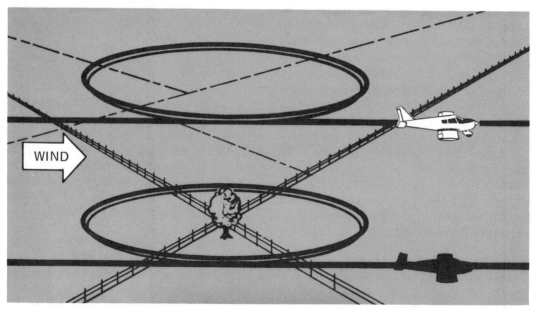

Fig. 6-20. Turns Around A Point

the center of an area *away* from livestock, buildings, or concentrations of people on the ground to prevent annoyance and undue hazard. The area selected also should afford the opportunity for a safe emergency landing.

FLYING THE MANEUVER

The easiest way to enter this maneuver is by flying *downwind* past the point at a distance equal to the radius of the desired turn. As the student arrives exactly *abeam* of the selected point, he should enter a medium bank turn toward the side on which his reference point is located. He then should carefully plan the track over the ground and vary his bank as necessary to maintain that track.

When any wind exists, it will be necessary to enter the steepest bank *immediately* since the airplane is headed directly downwind. Thereafter, the bank should gradually be shallowed until reaching the point where the airplane is headed directly upwind. At this position, the bank should gradually be steepened until the original bank angle is reached at the point of entry. The turn then should be continued in the same manner as described through another 360° of turn and the rollout made on the original entry heading. It will be seen that by entering on a heading directly downwind, the steepest bank is immediately established and all further maneuvering will involve shallower banks.

As the student gains experience, he will be required to fly turns about a point in *both directions.* Furthermore, he will be permitted to enter the maneuver *at any point.* The student should understand that the angle of bank at any given point is dependent upon the radius of the circle being flown, the wind velocity, and the groundspeed. When entering the maneuver at a point other than directly downwind, the angle of bank and the radius of the turn must be carefully selected in terms of wind velocity and groundspeed so that an excessively steep bank angle is not required later to maintain the proper ground track.

ACCEPTABLE PERFORMANCE

The student should be able to readily select the necessary ground references, maintain coordination, execute smooth, controlled turns, and maintain altitude within 100 feet of that selected. Excessively steep banks, flight below safe altitudes, and failure to employ proper collision avoidance techniques are disqualifying.

SECTION C—RECTANGULAR COURSES AND ELEMENTARY EIGHTS

RECTANGULAR COURSES

The objective of flying a rectangular course, or pattern as it is sometimes called, is to fly a definite ground track while maintaining an assigned altitude. Flying a rectangular pattern also teaches the methods used to track over the ground and the methods used to evaluate the angle of crab necessary to follow the desired track. (See Fig. 6-21.) Furthermore, practice in flying a rectangular course provides initial training in flying the airport traffic pattern. Another important objective is to learn the division of attention required to maintain the flight path, observe ground reference points, look for other traffic, and control the airplane while in the airport traffic pattern.

The instructor will select a field well away from other traffic where the sides are not more than a mile in length or less than one-half mile. These dimensions are rough guidelines, but the shape should be square or rectangular within the approximate limits given. The maneuver is initially flown at approximately 600

feet, and later at the altitude required for the airport traffic pattern. The bank angle in the turning portions of the pattern should *not exceed the medium bank* which is recommended for flight in the traffic pattern.

This maneuver requires the pilot to combine several flight techniques. First, a different crab angle may be necessary throughout the course, varied for each of the four legs. Second, he must track an imaginary line parallel to a fixed line. Third, the pilot must plan ahead and use different angles of bank in order to roll out of the turns at the proper distance from the field boundary as he flies around the corner of the rectangle. It may be of assistance to think of flying around each corner as performing *one-fourth* of a constant-radius turn, as shown in figure 6-21.

The flight path should not be *directly* over the edges of the field, but far enough away that the boundaries may easily be observed as the pilot looks out the side windows of the airplane. If an attempt is made to fly *over* the edge of

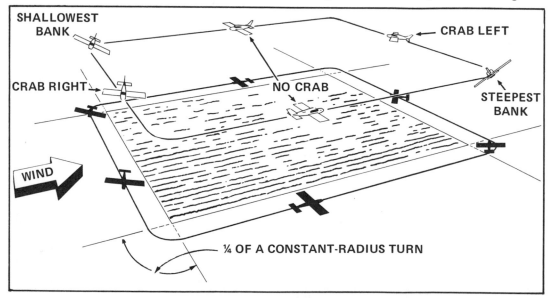

Fig. 6-21. Rectangular Course And Required Drift Corrections

the field, the required turns will be too steep or the maneuver will result in an elliptical, or egg-shaped course (somewhat defeating the objective of the rectangular pattern). The *closer* the track of the airplane is *to the edges* of the field, the *steeper* the bank required. The determining factor in deciding the distance from the field boundary to begin the maneuver should be based on the *normal distance* from the runway of the downwind leg of the typical airport traffic pattern.

When the proper position has been determined, the aircraft should be flown parallel to one side of the field until the corner is approached. A turn should then be started at the exact time that the airplane is *abeam* the corner so that a ground track *parallel to, and at the same distance from*, the next side of the field will be achieved upon recovery from the turn. This process is repeated at each corner and continued around the field for several trips. Again, the flight path should be such that the ground track of each resulting corner is *one-fourth* of a perfect circle.

After several circuits of the field, the direction of the flight path should be reversed and a rectangular course flown in the opposite direction for a similar number of trips. The student and his instructor will evaluate the number of trips around the rectangular course required to obtain proficiency; however, if the instructor specifies a certain number of circuits, it is the student's responsibility to count them and stop when the required number has been completed.

While flying the rectangular course, the student may also be requested to alternately glide on the simulated final approach leg and subsequently climb back to the simulated traffic pattern altitude in order to more realistically simulate the approach and climbout from an actual airport. The complexity of the maneuver is increased when a wind blows across the rectangular course *diagonally*, such as shown in figure 6-22. This is because a crab must be used *on all four legs* instead of *only on two legs*, as shown in figure 6-21.

The value of practicing rectangular courses should be obvious, since these procedures give the student experience in performing maneuvers similar to those required during actual takeoff and landing practice. Practice is devoted to these

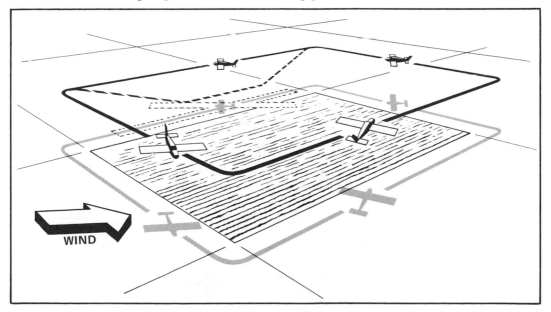

Fig. 6-22. Wind Blowing Diagonally Across Rectangular Course

exercises until fair proficiency is shown in all phases, correct ground tracks are being flown at the specified altitudes, and any traces of tension and confusion are eliminated. In summary, the principles used in flying rectangular patterns *apply directly* to airport traffic patterns, landing approaches, and departures.

ELEMENTARY EIGHTS

EIGHTS ALONG A ROAD

Practice in flying elementary eights often begins with *eights along a road.* In order to perform this maneuver, the student is instructed to select a road *parallel* to the wind and fly downwind above this road to a predetermined point, such as a fence row or the intersection of another road. At this point, he is instructed to execute a 180° turn using a medium bank and varying the bank angle to trace a half-circle ground track. (See Fig. 6-23, position 1.)

He then flies upwind along the road (position 2), remaining approximately the same distance from the road as when flying the rectangular course. The length of the straight-flight portion may vary from one-half mile to more than one mile.

After flying the straight-flight portion, another 180° turn should be made over a fenceline or road, as shown at position

3. This turn also should result in a perfect half-circle ground track and the roll-out should place the aircraft back over the road at position 4. Following this, the downwind leg is again flown and the other half of the "eight" is performed on the opposite side of the road in the same manner as described.

During the segment of the eight in which the pilot is headed downwind, relatively steeper banks will be required; however, shallower banks will be found necessary to counteract the effect of wind drift on the segments in which he is headed upwind. Of course, this principle is usually well established in the student's mind from previous practice of S-turns, turns around a point, and rectangular courses. Eights along a road are practiced in both directions until the proper compensation for wind in terms of changing bank angle and crab to maintain the desired ground track is achieved.

EIGHTS ACROSS A ROAD

Eights across a road or through an intersection are maneuvers that serve as a foundation for the slightly more difficult *eights around pylons.* The same principles of correcting for drift by varying the angle of bank, rate of turn, and crabbing the aircraft are required to fly this maneuver as in previous ground reference maneuvers.

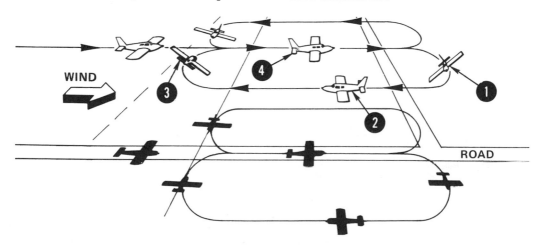

Fig. 6-23. Eights Along A Road

Eights across a road combine both turns and straight flight in order to maintain a preplanned path over the ground. (See Fig. 6-24.) The maneuver is usually performed at the intersection of two roads and, if possible, is flown when a moderate wind exists. The *objective* is to fly a loop on each side of the road intersection; each loop having the same radius. This requires careful planning as the bank *varies* from steep to shallow and back to steep again, as shown by the flight path in figure 6-24.

The four straight legs should be flown with the wings level, and the aircraft track should cross the intersection of the roads at the same angle each time, as shown in figure 6-24. Determining the proper time to roll into and out of banks will help the student establish the proper crab angle required and assist him in determining the length of the straight legs. These skills, plus varying the angle of bank to correct for wind drift, will further sharpen the pilot's planning skills and his ability to precisely maneuver the aircraft.

As a student becomes more proficient in flying the maneuver, he may be asked to perform the eights by rolling from one bank *directly* into the other while directly over the intersection. This increases the necessity for careful planning and speeds up the student's reactions. Proficiency in this maneuver provides a foundation for a more advanced maneuver called *eights around pylons.*

EIGHTS AROUND PYLONS

Eights around pylons are similar to eights across a road *except* the loops are made around two prominent points on the ground. Interestingly, in the early days of flight training, small towers, or *pylons*, were constructed to provide these reference points; hence, the name of the maneuver.

The entry to this maneuver should be made *downwind* perpendicularly toward a line drawn between the two points, as illustrated in the top portion of figure 6-25. The entry is made in a manner similar to the entry for turns around a point since the distance of the point of entry from the pylon and the wind velocity will determine the *initial angle of bank* required to maintain a *constant radius* from the pylon during each loop. The

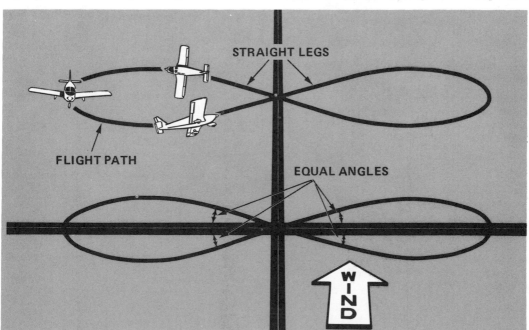

STRAIGHT LEGS

FLIGHT PATH

EQUAL ANGLES

WIND

Fig. 6-24. Eights Across A Road

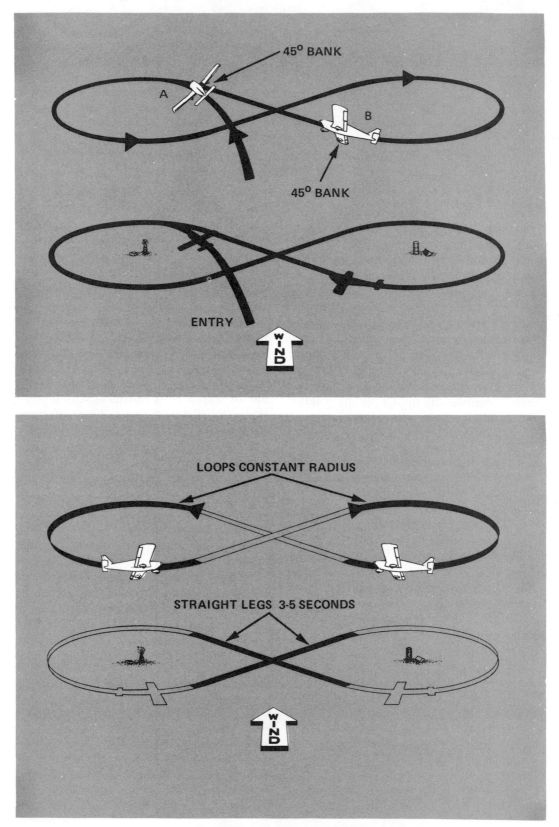

Fig. 6-25. Eights Around Pylons

airspeed should be set at normal cruise power and the maximum angle of bank should not exceed 45°.

As illustrated in the lower half of figure 6-25, the loops are made at a constant radius and both loops should be equal in size. The pylons should be selected and the turns planned so there is a *short duration of three to five seconds of straight flight* between the loops. This maneuver is another combination of two fundamental flight maneuvers — turns and straight-and-level flight.

Since the objective of the turning portions of eights around pylons is to maintain a constant radius from the pylon, this part of the maneuver closely resembles a turn around a point *except* that a complete circle is not made; therefore, all the previous discussions concerning turns about a point are equally applicable to the turning segments of eights around pylons.

BANK ANGLE

The amount of bank variation, obviously, will depend on the wind velocity as it does in S-turns, turns about a point, etc. If there is a wind, the bank will be continuously changing throughout the turns. The variation of the bank in each loop of the eight should be a gradual decrease from the steepest bank at turn entry to the shallowest bank as the airplane heads directly into the wind on the "ends" of the eight. At this point, a gradual increase in bank takes place until the steepest bank is again reached just prior to rollout for the straight-flight segment. Rollout from the turn should be completed on a heading that will insure that the flight during the straight-and-level segment will carry the aircraft to the turn entry point for the *second pylon*, where a turn of the same radius can be made around that pylon.

As the student gains experience in flying eights around pylons, he may be encouraged to start the maneuver at *any point* in order to save flying time and to improve his judgment and orientation. However, he should be sure that the direction of entry is such that the airplane will be flying downwind between the pylons, as shown in the lower portion of figure 6-25.

DRIFT CORRECTION

In a wind, the principles of "crab" drift correction must be employed in addition to varying the angle of bank in order to maintain the proper ground track around the pylon. Rollout from one loop may have to be made with the airplane *headed* toward some point on the upwind side of the next pylon. The exact heading is determined by trial and error. As a point from "which to work," the rollout from each loop should be made with the airplane headed toward the other pylon. An estimate of the amount of increase or decrease in the crab angle can be made on subsequent trials until the proper ground track is maintained.

ACCEPTABLE PERFORMANCE FOR GROUND REFERENCE MANEUVERS

Acceptable performance for ground reference maneuvers is evaluated on the basis of planning, orientation, drift correction, altitude control, and proper division of attention. The altitude should be maintained within 100 feet of an assigned altitude; the airspeed in *straight* flight should be within five knots of the speed specified for the maneuver, and in *turning* flight, within 10 knots. The loops of the eights should be symmetrical or of equal size. During the maneuvers, the aircraft must not be flown at an excessively steep bank angle or below a minimum safe altitude.

CHAPTER 7

ATTITUDE INSTRUMENT FLYING

PHYSIOLOGICAL FACTORS

A human being's ability to maintain equilibrium depends primarily upon three senses: the sense of sight, the sense of changing position which originates in the balance organs of the inner ear, and the postural sense which includes sensations of touch, pressure, and tension on muscles, joints, and tendons.

When flying in VFR conditions, the pilot's orientation is maintained by utilizing these same three senses. The primary sense is sight using the natural horizon of the earth as a reference. However, during instrument flight, the pilot must use the indications of his flight instruments to maintain his orientation with the earth.

During visual flight, some of the sensations from the inner ear and postural sense tend to send conflicting information to the brain. However, the sense of sight tends to provide sufficient information for the brain to make a correct interpretation of "what's going on" and tends to disregard the conflicting messages from other senses. When the instrument pilot's references based on sight are not available, sensory confusion may result as the brain attempts to interpret conflicting messages. Therefore, the pilot must learn to trust his instruments and react accordingly.

ATTITUDE INSTRUMENT FLYING

The term "attitude instrument flying" in itself implies a definition. Attitude instrument flying is control of an aircraft's position or attitude in space utilizing instrument reference rather than visual reference. In order to control the attitude, the pilot must become skillful at reading the six flight instruments shown in figure 7-1, interpreting the indications, and applying this information to control of the aircraft.

FUNDAMENTAL PILOT SKILLS

Instrument cross-check, instrument interpretation, and aircraft control are the three fundamental skills involved in all instrument flight maneuvers and are developed through training and practice. These skills may be developed individually, but as the pilot obtains proficiency, they are integrated into the unified, smooth, positive control responses required to maintain any prescribed flight path.

Fig. 7-1. Flight Instruments

CROSS-CHECK

Cross-checking, or scanning, is the term applied to the continuous systematic observation of flight instruments. The actual technique may vary somewhat with different individuals, various maneuvers, variation in aircraft equipment, and the experience and proficiency level of the pilot.

At first, without the guidance of the flight instructor, the student has a tendency to cross-check rapidly, looking directly at the instrument without knowing exactly what he is looking for. However, with direction, familiarity, and practice, the instrument scan will assume definite trends during specific flight conditions and these trends will be modified as the aircraft makes a transition from one flight condition to another. Also, the pilot learns that during some phases of flight the instrument scan must be more rapid than at other times.

Most individuals throughout their lifetime have learned to apply full concentration on a task at hand in order to perform the task well. However, this tendency toward complete concentration on one thing tends to introduce a number of common errors that can be eliminated through awareness of the erroneous tendencies and by conscious avoidance of the tendencies. Several common instrument scanning errors are listed as follows:

1. *Staring at a single instrument* — This tendency results from a natural human inclination to "watch this darn thing carefully and I'll get it right." Fixation on a single instrument usually results in poor control. For example, while performing a medium-bank or shallow-bank turn, the pilot may have a tendency to watch the turn indicator throughout the turn instead of including other instruments in his cross-check. This fixation on the turn coordinator would probably result in a loss of altitude through poor pitch and bank control.

2. *Instrument Emphasis* — Instead of relying on a combination of instruments necessary for airplane performance information, the student sometimes will place too much emphasis on a single instrument. This differs from fixation in that the student is using other instruments for information, but is relying too heavily on the information supplied by one particular instrument.

3. *Omission* — During the performance of a maneuver, the student sometimes fails to anticipate significant instrument indications following marked attitude changes. For example, during the level-off from a climb or from a descent, emphasis sometimes is placed on pitch control instruments while omitting the instruments which supply heading or roll information. Such omissions result in erratic control of heading and bank.

Lest the student be "scared out" before beginning, let it be stated that "it isn't all that hard." In fact, many students find that they can control the aircraft more easily and precisely by instruments. Occasionally, during visual flight, the instructor may admonish a student to "get his head out of the cockpit" in order to scan for other traffic and use outside visual references.

INSTRUMENT INTERPRETATION

The student should understand each instrument's operating principles, and this knowledge must be coupled with awareness of what the particular instrument reveals about the aircraft's performance. Each flight maneuver involves the use of *combinations* of instruments that must be read and interpreted in order to control the aircraft's attitude. For example, if pitch attitude is to be determined, the attitude indicator, airspeed indicator, altimeter, and vertical velocity indicator provide the required information. These instruments are enclosed in the shaded area shown in figure 7-2. If bank attitude is to be determined, the attitude indicator, the turn coordinator, and the heading indicator must be interpreted. (See Fig. 7-3.) It can be seen, then, that all the flight instruments must be included in the scan during flight by instrument reference.

Once the techniques of instrument scan and instrument interpretation reach an elementary learning level, these techniques may be used to obtain the final objective, aircraft control.

AIRCRAFT CONTROL

Even though instrument indications have been substituted for outside visual references, the aircraft is controlled in the same manner as when flying by visual reference. Specifically, the student must learn to *substitute* the *artificial horizon*, or attitude indicator, *for the natural horizon*. He must also *substitute* the heading indicator, or *directional gyro*, for a visual reference point on the horizon.

To control an aircraft in flight by reference to instruments, the pilot must continue perfecting the techniques of proper pitch, bank, and power control as practiced during flight by visual reference. The pilot should endeavor to maintain a "light touch" on the controls and trim off any control pressures once the aircraft has stabilized in a particular attitude. Abrupt and erratic aircraft control and pilot fatigue result, if light control pressures are not utilized and control pressures are not relieved with trim.

STRAIGHT-AND-LEVEL FLIGHT

Since the basic elements of straight-and–level flight were presented in chapter 2, the two integral components, pitch control and bank control, will be *reviewed* herein.

Fig. 7-2. Pitch Instruments

Fig. 7-3. Instrument Clues to Bank Attitude

PITCH CONTROL

At a constant airspeed and power setting, there is *only one* specific pitch attitude which will maintain level flight. The instruments used to maintain pitch control are the attitude indicator, airspeed indicator, altimeter, and vertical velocity indicator. Any change in the pitching motion of the aircraft registers a change on these instruments in *direct proportion* to the magnitude of the pitch change.

In a climb, as shown in figure 7-4, the nose of the small airplane rises above the artificial horizon bar of the attitude indicator. This causes a decrease in airspeed, an increase in altitude, and a positive rate-of-climb indication.

Conversely, a pitch change to a nose—down attitude causes the small airplane to move below the horizon bar on the attitude indicator. In addition, an increase in airspeed, a decrease in altitude, and an indication of a descent on the vertical velocity indicator are displayed on the flight instruments. (See Fig. 7-5.)

ATTITUDE INDICATOR (PITCH)

The attitude indicator is the one instrument on the panel which, by itself, gives a pictorial display of the aircraft's overall attitude. All of the other instruments are used indirectly in determining attitude.

The relationship between the natural horizon and the artificial horizon bar during pitch changes is often demonstrated by the instructor during instrument flight instruction. From this demonstration, the student will observe that the displacement of the horizon bar on the attitude indicator appears to be much *smaller* than the displacement of the aircraft nose in relationship to the natural horizon. Therefore, "fingertip" control pressures should be used and the required attitude changes made slowly and smoothly.

In practicing pitch control using the attitude indicator, the student should initially *restrict* the displacement of the horizon bar to a *half-bar width* up or down, progressing later to a full-bar

Fig. 7-4. Nose-up Instrument Indications

Fig. 7-5. Nose-down Instrument Indications

width. Greater displacement should be used only when significant attitude changes are required.

The control pressures necessary to effect pitch changes will vary in different aircraft. However, the student cannot feel control pressure changes if he has assumed the "white knuckles" grip on the control wheel. If light pressures are maintained and the aircraft retrimmed after it is stabilized in a new attitude, a smooth and precise attitude control technique will result.

ALTIMETER

At a constant airspeed and power setting, any deviation from level flight will be the result of a pitch change. The altimeter provides an *indirect* indication of the pitch attitude. Since the altitude should remain *constant* when the aircraft is in level flight, a deviation from the desired altitude indicates the need for a pitch change.

Obviously, if the altitude is increasing, the nose must be lowered. The *rate* of movement of the altimeter needle is as important as its *direction* of movement in maintaining level flight. Large pitch attitude deviations from level flight result in rapid altitude changes; slight pitch deviations produce much slower changes in the altimeter needle movement.

ALTITUDE CORRECTION RULES

A common rule for altitude corrections of *less than 100 feet* is to use a *half-horizon-bar width* adjustment on the attitude indicator; for corrections in *excess of 100 feet*, a full-bar width correction should be used. As these corrections are established, the rate of altitude change on the vertical speed indicator and the altimeter may be observed in the instrument scan.

Another rule of thumb may be stated as follows: if the deviation from the desired altitude is *less than 100 feet*, the attitude adjustment needed to return to the correct altitude may be made *without* changing the power setting; if the deviation from the desired altitude is *greater than 100 feet*, a change in power setting as well as an appropriate compensation in pitch attitude should be performed.

VERTICAL VELOCITY INDICATOR (VVI)

The normal function of the vertical velocity inidcator is to aid in establishing and maintaining a desired rate of climb or descent. Due to the design of the instrument, there is approximately a six to nine-second lag before the correct *rate* of change is registered. Even though this lag exists, the VVI may be used as a *trend* instrument for maintaining a desired pitch attitude. As a trend instrument, it indicates the direction of pitch change *almost instantaneously*. (See Fig. 7-6.)

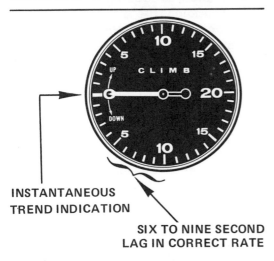

INSTANTANEOUS
TREND INDICATION

SIX TO NINE SECOND
LAG IN CORRECT RATE

Fig. 7-6. Vertical Velocity Instrument

If the VVI needle is observed to deviate from the *zero* position, the instrument is indicating that a pitch change *is in progress*. As the pilot applies corrective pressures while referring to the attitude indicator, this pressure will stop the needle movement and place the aircraft in a level attitude again. However, the student is cautioned not to try to *return the needle* to zero since the lag in the vertical velocity indicator will induce a tendency to overcontrol. Rather, after a pitch correction has been made to return to level flight attitude, an additional slight pitch correction should be made to return to the desired altitude. In sum-

mary, the vertical velocity indicator is combined in the cross-check with the attitude indicator and the altimeter to maintain level flight.

AIRSPEED INDICATOR

The airspeed indicator, when properly interpreted, represents another *indirect* indication of pitch attitude. If a constant power setting and pitch attitude is established and the aircraft permitted to stabilize, the airspeed will remain constant. As the pitch attitude is raised, the airspeed *decreases* slightly. On the other hand, as the pitch attitude lowers, the airspeed *increases* somewhat.

A *rapid change* in airspeed is an indication that a *large pitch change* has occurred and that smooth control pressure in the opposite direction should be applied. Again, however, the amount of pitch change caused by the control pressure should be noted on the attitude indicator to avoid overcontrolling the aircraft. If the airspeed indicator needle is moving in one direction, the instant at which it hesitates is the time at which the aircraft is passing through the level flight attitude. Therefore, the airspeed indicator, when included in the scan with the attitude indicator, altimeter, and vertical speed indicator, furnishes positive pitch control information to the pilot.

BANK CONTROL INSTRUMENTS

If the individual spends some time in the training environment around a typical airport, one of the commonsense comments frequently heard from flight instructors is, "If the wings are kept level, the aircraft will not turn." Therefore, in order to fly a given heading, the pilot should attempt to keep the wings of the aircraft level with the horizon while maintaining coordinated flight.

ATTITUDE INDICATOR

The principal instrument used for bank control is the attitude indicator. It is supported in this function by the heading indicator and turn coordinator. The

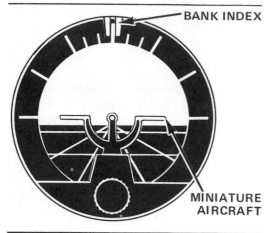

Fig. 7-7. Attitude Indicator

student will recall that on the standard attitude indicator, the angle of bank is shown by the alignment of the bank index with the bank scale at the top of the instrument and by the relationship of the wings of the miniature aircraft to the artificial horizon, as illustrated in figure 7-7.

HEADING INDICATOR

The bank attitude of an aircraft in coordinated flight also is *indirectly* indicated on the heading indicator. Generally, if the heading displayed on the heading indicator is not changing, the wings are level. On the other hand, a slow heading change would be indicative of a shallow bank angle, and a rapid change in heading would indicate a steep bank.

TURN COORDINATOR

When the small airplane silhouette in the turn coordinator is in a wings-level position, the aircraft is maintaining a *constant heading*. If the wings of the small airplane are displaced from the level-flight position, *the aircraft is turning* in the direction that the small airplane is banking. The turn coordinator not only indicates the direction of bank in a turn, but also the *rate* of turn. As explained in chapter 2, when a wingtip of the minia-

ture airplane is aligned with one of the white marks at the side of the instrument below the level-attitude mark, the airplane is turning at the standard rate of three degress per second. (See Fig. 7-8.)

If deviation from the desired attitude is observed on any of the bank control instruments, the correction should be made in the two-step method outlined previously. First, stop whatever is happening; and second, apply an appropriate correction to return to the desired attitude.

In making bank corrections, the principal reference is the attitude indicator. A rule of thumb for making the corrections is given as follows: to return to a desired heading, use a bank angle equal to *one-half of the difference* between the present heading and the desired heading, but in no case use a bank angle greater than will produce a standard rate turn. For example, if the desired aircraft heading is 300° and the present heading is 290°, the required heading change is 10°. Therefore, the bank angle used to return to the desired heading should be no greater than one-half of 10° or, in this example, 5° of bank.

A Few Words About The "Ball"

The ball of the turn coordinator is actually a separate instrument, although

Fig. 7-8. Standard Rate-of-Turn Indication

the ball and the small airplane silhouette are used together. It should be recalled that the ball indicates whether coordinated flight is being maintained. If the ball is off center, as shown in figure 7-9, the aircraft is slipping and correction should be made with appropriate coordinated rudder and aileron pressure.

Fig. 7-9. Turn Coordinator
Displaying A Slipping Condition

CLIMBS

To enter a climb from cruising airspeed, the nose should be raised to a two-bar width, nose-high pitch attitude, as indicated in figure 7-10. This recommended pitch attitude adjustment is a "rule of thumb" and will vary with the aircraft used, the desired rate of climb, and the desired climb airspeed.

Light control pressures should be used to initiate and maintain the climb since pressures will change as the aircraft decelerates. These pressures may be more easily detected and trimmed off if the pilot maintains a light touch on the controls. Power may be advanced to the climb power setting simultaneously with the pitch change (if maximum r.p.m. will not be exceeded), or power may be added after the pitch change is established and the airspeed approaches the climb speed.

If the transition has been made smoothly, the vertical speed indicator will show an immediate upward trend and will stabilize at a rate appropriate to the stabilized airspeed and pitch attitude. As the transition is made from cruise to climb, additional right rudder pressure must be added in order to maintain the desired heading.

Directional control is maintained during a climb entry by reference to the attitude indicator, heading indicator, and turn coordinator. Once the climb has been established and stabilized, control pressures should be trimmed off.

The typical instrument indications of a training aircraft in a stabilized straight climb are shown in figure 7-11. If the instrument scan should reveal any deviation from the desired heading or pitch attitude, the required correction is made by reference to the attitude indicator while continuing the scan.

LEVEL-OFF FROM A CLIMB

To level off from a climb at a designated altitude, the student should initiate the level-off *before* reaching the desired altitude. The aircraft will continue to climb at a decreasing rate throughout the transition to level flight. An effective "rule

Fig. 7-10. Initial Climb Pitch Attitude

Fig. 7-11. Stabilized Climb Indications

of thumb" is to *lead the altitude by 10 percent of the vertical velocity indication.* For example, if the aircraft is climbing at 500 feet per minute, the student should begin the level-off 50 feet prior to reaching the desired altitude.

To level off and accelerate to cruising speed, smooth, steady, forward elevator pressure should be applied. As the level flight attitude is established using the attitude indicator, the airspeed will increase to cruising speed. Since forward control pressure must increasingly be applied, appropriate "rough trim" may be made during the transition. Power reduction to the recommended cruise setting is made as cruise speed is attained. As the level flight attitude is established, the vertical speed needle will move slowly toward zero. Also, the altimeter needle should slow its upward movement as the airspeed indicator needle moves toward the desired cruise speed.

During the transition from climb to cruise flight, a simultaneous relaxation

of right rudder pressure (if not trimmed off during the climb) should be made and the heading indicator scanned to maintain the desired heading. As the desired airspeed is achieved, final "fine trim" should be made.

DESCENTS

To enter a descent, the pilot reduces the airspeed to a predetermined descent airspeed using pitch and power adjustments while in straight-and-level flight. The descent attitude then is established using the attitude indicator and the power settings are adjusted to a value that result in the approximate desired speed. As the pitch attitude and descent rate stabilize, the aircraft is trimmed.

During a constant airspeed descent, any deviation from the desired airspeed will require a pitch adjustment. For a constant *rate* descent at a specified airspeed, the entry is the same but the vertical velocity indicator becomes the principal instrument for pitch control as it stabilizes at the desired rate. Again, airspeed adjustments should be made primarily with changes in pitch attitude and the rate of descent corrections should be made by changing the power setting. Figure 7-12 illustrates the instrument indications representative of a stabilized descent in a typical light training aircraft.

LEVEL–OFF FROM A DESCENT

As in other level-off procedures, the pilot must again begin leveling off from a descent *prior* to reaching the desired altitude; the amount of lead depends primarily upon the *rate of descent*. Assuming a standard 500-foot per minute rate of descent, leading the level-off altitude by 100 to 150 feet will provide an initial guideline. As the lead point is reached, power is added to the appropriate level-flight cruise setting. Since the nose will tend to pitch up as the airspeed increases, forward elevator pressure must be applied in order to maintain the descent until approximately 50 feet above the desired altitude. At this point, the pitch attitude should be adjusted

smoothly to level-flight attitude for the airspeed selected.

The attitude indicator must be included frequently in the scan as the level-off procedure is accomplished. After the pitch attitude and the airspeed have stabilized in straight-and-level flight, control pressures should be removed by trimming.

TURNS

To enter a turn, the pilot should apply coordinated aileron and rudder pressure in the desired direction of turn. The attitude indicator is used to establish the approximate angle of bank (based on true airspeed) required for a standard rate turn. The approximate angle of bank is achieved by placing the bank angle mark within the white paired indices at the top of the attitude indicator, as shown in figure 7-13.

As the bank is established, additional lift is required to offset the loss of vertical lift component in the turn, so the nose

Fig. 7-12. Instrument Indications In A Stabilized Descent

must be raised slightly to maintain altitude. (See Fig. 7-13.) Therefore, as the turn is established, it is necessary to adjust the nose of the miniature airplane slightly above the level flight position on the horizon bar. As the turn is progressing, the turn coordinator is checked to determine if a standard rate is being maintained and, if not, a bank adjustment is made on the attitude indicator.

The pilot also must include the heading indicator in his scan to determine progress toward the desired heading. Furthermore, the altimeter is checked to determine that the adjusted pitch attitude has properly compensated for the loss of vertical lift component and that a constant altitude is being maintained throughout the turn.

Approximately eight degrees before reaching the desired heading, the pilot should apply coordinated rudder and aileron pressure to roll out of the turn.

His principal instrument reference for the rollout should be the attitude indicator. Since a slightly nose-high attitude has been held throughout the turn, back pressure on the control column must be relaxed to prevent an altitude gain as the aircraft is returned to straight-and-level flight. As the wings-level position is attained, the pilot must continue his instrument scan. As the aircraft stabilizes in the cruise flight configuration, any control pressures required to maintain straight-and-level flight should be trimmed off.

A rule of thumb in determining the amount of lead required for rollout from a standard rate turn is to use approximately one-half the angle of bank. For example, if the standard rate turn is being made at a bank angle of approximately 15°, the student should begin the rollout approximately eight degrees before reaching the desired heading. (See Fig. 7-14.)

Fig. 7-13. Level Standard Rate Turn Instrument Indications

Fig. 7-14. Leading The Rollout

THE PROPER EXECUTION OF CLIMBING AND DESCENDING TURNS

The proper execution of climbing and descending turns combines the techniques used in straight climbs and descents with turning techniques. Initially, the climb or descent is established, and as the pitch attitude is stabilized, the pilot rolls into the turn. However, proficiency should be developed so that the climb or descent can be established simultaneously with the turn. The pilot must carefully consider aerodynamic factors that affect lift and power control; the rate of cross-check and interpretation must be increased to enable him to control the bank as well as pitch changes.

CRITICAL ATTITUDES AND RECOVERIES

Any aircraft attitude not normally used in the course of flight is considered an *critical attitude*. Such an attitude may result from any number of conditions; for example, turbulence, disorientation, confusion, preoccupation with cockpit duties, carelessness in cross-check, errors in instrument interpretation, or lack of proficiency in basic aircraft control.

By the time the pilot realizes the need to "get on the gauges," the attitude of his aircraft may be such that immediate attention and recovery are required.

During the more advanced phases of private pilot training, the pilot may be instructed to take his hands and feet off the controls and close his eyes. The instructor will then put the aircraft in a critical attitude. The attitude most likely will be an approach to a climbing stall or a well-developed power-on spiral, as recovery from these particular maneuvers are specifically required on the FAA flight test.

The student should expect that his flight instructor will maneuver the airplane through a number of preliminary slight climbs, descents, slips, etc., and may require the student to move his head up and down in order to induce disorientation. As soon as the flight instructor says, "You've got it," the pilot should immediately check the airspeed to determine if it is increasing or decreasing. *In all cases*, recoveries from critical attitudes are made to straight-and-level flight.

RECOVERY FROM THE APPROACH-TO-A-CLIMBING STALL

If an approach-to-a-climbing stall is indicated by an airspeed indication that is low and decreasing, as shown in figure 7-15, the following recovery procedures should be used:

1. Apply control pressures to lower the nose toward level flight attitude.

2. Immediately add full power.

3. Simultaneously level the wings.

4. Reestablish straight-and-level flight, permit the airplane to reach cruise airspeed, reduce power to cruise setting, and properly trim the aircraft.

RECOVERY FROM A WELL – DEVELOPED POWER – ON SPIRAL

The well-developed, power-on spiral is the flight condition normally encountered *by a noninstrument rated pilot continuing flight into adverse weather conditions.* In this situation, the pilot senses the increased speed of the aircraft due to the increased slipstream noise and obviously high engine r.p.m. The normal reaction to a nose-down attitude is to apply back pressure to the control column; however, in this instance this will result only in a continually tightening spiral or, as it is sometimes termed, a "graveyard spiral."

Therefore, if the airspeed is observed to be high and increasing, and the engine and slipstream noises are as described, the following recovery procedure is recommended:

1. *Reduce power* to prevent excessive airspeed and further loss of altitude.

2. Correct the bank attitude to wings level *before* any application of back pressure.

3. Return the pitch attitude to straight-and-level flight and, as the aircraft decelerates, reestablish the cruise power setting.

Most new training aircraft are equipped with nontumbling gyros; however, when the pitch and bank limits of some older gyros are exceeded, as is possible during the well-developed, power-on spiral, the attitude gyro may be completely unreliable in effecting the pitch and bank adjustments necessary for the recovery. In this case, immediately after power is reduced the turn coordinator or turn needle is observed to determine the direction of turn; then the aircraft is rolled in the opposite direction until the turn has stopped. At that point, back pressure may be applied to the control column until the aircraft has again

Fig. 7-15. Approach-To-A-Climbing Stall Indications

Fig. 7-16. Power On Spiral Indications

attained a level flight attitude, as indicated by the attitude indicator, airspeed indicator, vertical velocity indicator, and altimeter. The instrument indications in a typical, well-developed, power-on spiral are shown in figure 7-16.

ACCEPTABLE PERFORMANCE

Performance will be evaluated on smoothness, coordination, and accuracy. Turns of more than 180° duration should be made with recovery to within 20° of a preselected heading and 100 feet of a preselected altitude. The student must be able to control the aircraft solely by reference to flight instruments without entering a stall or exceeding the operating limitations.

A FEW WORDS ABOUT THESE NEWLY ACQUIRED SKILLS

The basic elements of attitude instrument flying are presented for the non–instrument rated pilot, *not* to qualify him to fly in instrument conditions, but rather to enable him to "get the heck out of a situation that he shouldn't have gotten into in the first place." Even after the pilot has attained reasonable skill in performing the basic instrument maneuvers and recoveries from critical attitudes listed in this chapter, he should *never* undertake flight in instrument conditions. However, these newly acquired skills will serve as a firm foundation for further instrument training and the eventual acquisition of an instrument rating.

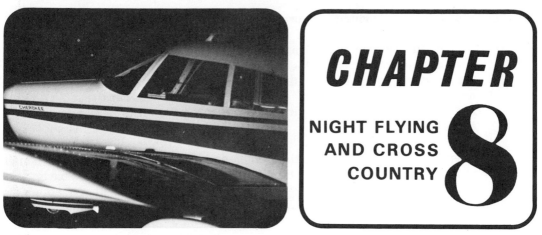

SECTION A—NIGHT FLYING

INTRODUCTION

It was not too many years ago that most general aviation pilots considered night flying an emergency procedure. Since then, great strides have been made in navigation, lighting, marking, radio communication, etc. which have made night flying in light aircraft a routine operation. Furthermore, airports, obstructions, and aircraft are now equipped with standard lighting, greatly simplifying the task of the pilot flying at night.

ADVANTAGES OF NIGHT FLIGHT

Perhaps one of the greatest advantages of night flight can be realized by a comparison of the loss of utility a person would experience if he operated his automobile only in the daytime. Since the airplane is also a transportation vehicle, the ability to operate safely after dark greatly extends the *utility* of the aircraft.

In many respects, night flight is easier and more pleasant than daytime flying. Traffic is usually easier to spot at night, the air is generally smoother and cooler, resulting in more comfortable flight and better airplane performance. Furthermore, the night pilot experiences less airport traffic pattern congestion and often finds less competition when using the various communication frequencies.

The aesthetic advantages of night flight should not be neglected, since most pilots feel more detached from earthly hustle, bustle, and care. This is especially true on a smooth, clear night when the pilot can turn the cabin lights down, relax, and be drawn closer than ever to the magic and mystery of flight as he watches the airplane wings slide across the sparkling lights of a city, or he visually traces the "strings of glittering beads" of automobile headlights fading into the distance.

THE "GO, NO-GO" DECISION

On a silvery bright moonlit evening when the visibility is good and the wind is calm, night flying is a real delight and not a great deal different than flying during the day. However, when the wind is gusty and the temperature/dewpoint spread is small, and possibly the landing light doesn't work, the complexion of the situation changes quite radically. Therefore, the inexperienced pilot must carefully consider his "go, no-go" decision in light of the following factors:

1. visibility
2. amount of ambient light (moonlight, city lights, etc.) available
3. surface winds
4. general weather situation

5. availability of lighted airports enroute
6. proper functioning of the aircraft and its systems
7. night flying equipment in the aircraft
8. the pilot's recent night flying experience

These concepts will be developed in the following pages in order to assist the pilot in making sound decisions when considering flying after the sun goes down.

PREFLIGHT INSPECTION

The preflight inspection should be performed in the usual manner, preferably in a well-lighted area with the aid of a flashlight. In addition to being valuable during the preflight inspection, a readily available flashlight becomes very important in case of a malfunction of any of the instrument or cabin lights. Some pilots prefer to place a layer of red cellophane between the flashlight bulb and its lens in order to provide red light which is less detrimental to night vision. A very important additional preflight inspection procedure is insuring that a spare set of fuses is aboard the aircraft.

During daylight operation, a clean windshield and side windows are certainly assets. At night, windshield cleanliness is imperative, since cigarette smoke, smudges, etc. interfere with vision to a greater degree than during daytime.

POSITION LIGHTS

All aircraft operating between sunset and sunrise are required to have operable navigation lights, sometimes called *position lights*. These lights should be turned on prior to the airplane walk-around so that they may be visually checked to insure proper operation. Many flight instructors prefer that the position lights be checked immediately and then turned off to avoid excessive drain on the bat-

tery. Position lights should be in the ON position anytime the engine is running or when the aircraft is moving.

The position lights required by regulations are a *red* light on the left wingtip, a *white* light on the tail, and a *green* light on the right wingtip, as shown in figure 8-1. Additionally, many aircraft include running light detectors - small plastic attachments to the navigation lights which convey light up above or below the surface of the wing and thus allow the pilot, observing from the cabin, to see if the light is operating.

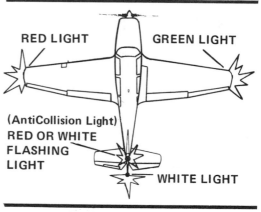

RED LIGHT   GREEN LIGHT

(AntiCollision Light)
RED OR WHITE FLASHING LIGHT

WHITE LIGHT

Fig. 8-1. Position Lights

LANDING AND TAXI LIGHTS

Most modern training aircraft include a landing light and possibly a taxi light. The only difference between the two is that the taxi light is mounted so as to illuminate the taxiway immediately ahead of the aircraft. Conversely, the landing light is aimed at a higher angle and designed to illuminate the runway well ahead of the landing aircraft. These lights may be mounted behind a common lens in the leading edge of the wing, as shown in figure 8-2, or possibly in the cowling, as illustrated in figure 8-3. Another installation usually restricted to higher performance airplanes is described as a retractable gear mounted light, as shown in figure 8-4.

Although both the landing light and the taxi light must be visually checked for

Fig. 8-2. Landing And Taxi Lights Mounted
In The Wing

Fig. 8-3. Cowing-Mounted Lights

Fig. 8-4. Gear-Mounted Lights

clude inspection for illumination, cracks in the lens, and the correct aiming angle of each installed light. It is considered, however, very discourteous to activate the taxi and landing lights when they are directed at another aircraft which may be taxiing or landing since this will contribute to loss of night adaption of the oncoming pilot.

ANTICOLLISION LIGHTING

All recently manufactured aircraft certified for night flight must incorporate anticollision lights which render an aircraft more visible to other pilots at night. The most common type is a flashing or rotating anticollision beacon which normally emits red flashes of light at a rate of approximately one flash per second.

An increasing number of airplanes, however, are being equipped with brilliant flashing white strobe lights which can be seen for many miles at night. In fact, strobe lights are sufficiently bright to remain useful even in poor visibility conditions during the daytime. If installed, optional lighting system(s) should also be checked during the preflight.

INSTRUMENT PANEL LIGHTS

All modern aircraft are equipped with some system for lighting the instrument panel and the instruments. Prior to any night flight, the panel lighting system should be checked to determine that it is operating satisfactorily. Panel lighting is generally controlled by a rheostat switch. This allows the pilot to select the intensity of light which best satisfies his needs. Normally, there is a separate rheostat for the flight instruments, engine instruments, and radios.

The lighting intensity should be adjusted just bright enough so the pilot is able to read the instrument indications. If the lighting is too bright, a glare results and night vision suffers. Three types of panel lighting are found in today's aircraft.

FLOOD LIGHTING

Flood lighting is a common method which illuminates the entire instrument

correct operation during the airplane walk-around, these lights should not be allowed to operate for any length of time (with the engine shut down) due to the high electrical energy drain on the battery. The preflight check should in-panel with one light source. In this system, a single ceiling-mounted light is used with a rheostat to regulate its intensity. Its beam is directed over both the flight and engine instruments. Flood lighting seems to produce the most glare if the intensity is too high.

POST LIGHTING

Post lighting is found in slightly higher performance aircraft. Each instrument has its individual light source, which is located on a small, protruding "post" adjacent to the instrument. Each post light beam is directed at the instrument face and shaded from the pilot's eyes.

Generally, there are two rheostats for this lighting system; one for the flight instruments and the other for the engine instruments. In addition, there may be separate controls for illumination of fuel tank selectors, switch panels, radio gear, and convenience lighting.

INTERNAL LIGHTING

Internal instrument lighting is similar to post lighting, except for the fact that the light source is located *inside* the instrument itself. Again, the intensity of light is usually controlled in two groups—flight and engine instruments.

The magnetic compass and radios generally utilize internal lighting. This type of lighting normally produces the least amount of glare and is found in more advanced aircraft.

NIGHT VISION

The ability to see at night can be greatly improved by understanding and applying certain techniques. The pilot's night visual ability can actually be increased by practice. If the pilot's eyes are exposed to strong light, even briefly, night vision is temporarily destroyed. For this reason, avoidance of strong light must begin well in advance of a night flight since several

minutes are required to regain night vision.

It has been found that dark adaptation (night vision) is destroyed most quickly and completely by exposure to white light, while dark red light has been found to be the least detrimental. Although red light is most desirable for preservation of dark adaptation, its use results in disturbance of normal color relationships. Therefore, carefully designed systems of white or blue-white light are finding wider application for cabin illumination.

OFF CENTER VISION

When viewing an object, we normally look directly at it utilizing central vision in order to see it clearly; however, central vision is ineffective under low illumination. For this reason, the pilot should not look directly at objects at night. The objects can be seen clearer if the gaze is directed slightly above, below, or to one side of the object. It has been found that looking about 10° off center will permit better viewing during low levels of lighting.

The above effect can be demonstrated by counting a cluster of very faint lights in the distance at night. When looking slightly off center, more lights will be clearly seen than when looking directly at the cluster.

CABIN FAMILIARIZATION

Following the preflight inspection, one of the first steps that every pilot should take in preparation for night flight is to become *thoroughly familiar* with the aircraft cabin, instrumentation, and control layout. It is recommended that practice in locating each instrument, control, and switch be made. The pilot should be able to do this in the dark as well as with the instrument lights on. Since the markings on some switches and circuit breaker detection devices may be hard to read at night, the pilot should assure himself that he is able to locate, use these devices, and read the markings in poor lighting conditions.

Many instructors prefer to give the student what is commonly termed "a blindfold cockpit check." Generally, this is not actually accomplished with a blindfold; however, the student is instructed to close his eyes. Then the instructor will state, for example, "Where is the circuit breaker for the primary aircraft radio?" Without opening his eyes, the student should be able to place his fingers on the required item and, in this particular example, be able to reset the circuit breaker without opening his eyes. This type of practice is continued between student and instructor until the student is thoroughly familiar with the location of all of the controls, instruments, switches, and markings and can locate them without the aid of sight.

AIRPORT LIGHTING

ROTATING BEACONS

Most airports, including many of the smallest, are equipped with rotating beacons to make them easier to locate at night. Beacon-equipped civil airports for land planes produce alternating green and white flashes, as shown in figure 8-5. Military airports can be distinguished from civil airports because their beacons emit *two white flashes* alternating with a single green flash. (See Fig. 8-6.)

TAXIWAYS AND RUNWAYS

The painted markings on runways, ramps, and taxiways are not especially useful to pilots at night since they are difficult to see. Therefore, various types of lighting aids are used to mark and identify different segments of the airport for night operations. (See Fig. 8-7.)

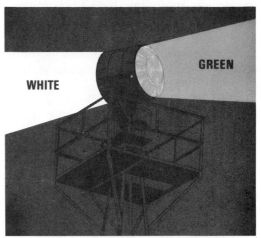

Fig. 8-5. Beacon—Civil Land Airport

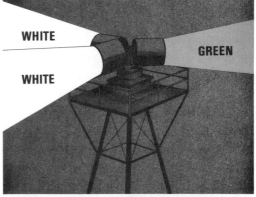

Fig. 8-6. Beacon—Military Land Airport

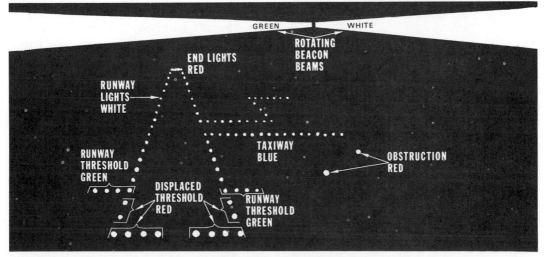

Fig. 8-7. Airport Lighting Aids

Taxiways are marked along their edges with blue lights to distinguish them from runways, which have yellow/white lights along the edges. The intensity of taxiway and runway lights can be controlled by the tower and may be adjusted upon request of the pilot. The threshold of a runway is marked with two or more green lights and obstructions or unusable areas are marked with red lights, as illustrated in figure 8-7.

OTHER LIGHTING AIDS

Wind indicators such as wind socks, wind tees, and wind tetrahedrons are often lighted so that they are clearly visible from the surface and from traffic pattern altitude. (See Fig. 8-8.) Also, radio voice communication, when available, should be considered an effective supplement to obtain surface wind and active runway information at night.

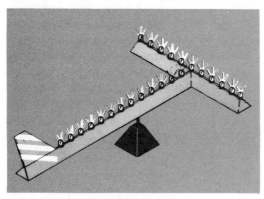

Fig. 8-8. Lighted Wind Tetrahedron

ENGINE STARTUP

Caution should be used in the engine startup procedure at night since it is difficult for other persons to determine that the pilot *intends* to start the engine. In addition to the usual "clear prop," turning on the position lights or momentarily turning on the landing light may help warn others that the propeller is about to rotate. An alternate procedure, especially useful in cold weather where conservation of electrical energy is especially important, is to shine the flashlight through the windshield and around

the area of the propeller prior to starting the engine.

TAXI TECHNIQUE

After the engine is started (oil pressure checked, etc.), other necessary electrical equipment should be turned on, but the taxi light left off until the pilot is actually ready to taxi. Aircraft taxi and landing lights normally do not cast the wide beam illumination characteristic of automobiles, but rather the beam tends to be more narrow and concentrated. Because of this fact, taxi light illumination to the side is *minimal* and taxi speed should be slower at night, especially in congested ramp and parking areas. Initially, judgment of distances is more difficult and it takes some adaptation to taxi within the limitations of the area covered by the taxi light.

When operating at an unfamiliar airport at night, it is wise to ask for instructions or advice about taxi procedures since it is embarrassingly easy to wander into rough unpaved areas, areas of construction, or unlighted and unmarked parts of the ramp and taxiway. Ground controllers or UNICOM operators are usually very cooperative in furnishing pilots with this type of information.

RUNUP

Once the runup position is reached and the aircraft stopped, the taxi or landing lights and any unnecessary electrical equipment should be turned off to conserve electrical power until the runup is completed. In addition to the usual runup procedures, the radios should be carefully checked for operation and, during the runup, the pilot should watch for a drop in the intensity of the lighting equipment when power is reduced to idle. A pronounced drop in intensity may be a clue that the battery needs charging; although if the battery is badly discharged, the pilot probably would not have been able to start the engine with the starter in the first place.

TAKEOFF

Many flight instructors minimize the differences between day and night flight by scheduling a night checkout which *begins at twilight*. When this procedure is utilized the takeoffs, landings, and traffic pattern work begin in the more familiar daylight environment and, as darkness increases, the transition into the conditions associated with night operation is made gradually. *This procedure is highly recommended.*

Landing lights are *normally* used during the takeoff roll. However, after experience is gained, the pilot should make several takeoffs *without* the aid of the landing lights. As the takeoff roll is initiated, the pilot should use a reference point down the runway or, in other words, the point where the runway boundary lights seem to converge.

He will find that the runway boundary lights (seen *in the peripheral vision)* assist in keeping the airplane correctly aligned during the takeoff roll. The initial tendancy of many pilots is an illusion of great speed because of the inclination to look *directly* toward the more obvious references, the boundary lights rushing alongside the aircraft.

Although there is nothing particularly difficult about taking off at night, the first takeoff may present a rather striking visual impression due to the lack of reliable outside visual references after the pilot is airborne. This is particularly true at smaller airports located in sparsely populated areas.

The recommended procedure is to utilize the flight instruments as well as available outside visual references immediately after takeoff. Immediately after the lift-off, the usual ground references will disappear and the pilot should maintain a normal climb pitch attitude on the attitude indicator. The vertical speed indicator and altimeter should register a

climb and the airspeed indicator should also be included in the cross-check. The first 500 feet of altitude after takeoff is considered to be the critical period of time in *transitioning* from the comparatively well-lighted airport area into what sometimes appears as a "black hole in the night."

VISUAL IMPRESSIONS

Most fledgling night pilots find the initial visual impressions after traffic pattern departure to be strikingly different than those they are accustomed to during daytime flying. Therefore, orientation in the local familiar flying area helps the pilot relate chart information and actual terrain and landmarks under night conditions.

The outlines of major cities and towns are clearly discernible and major metropolitan areas are visible (during favorable weather) from distances up to 100 miles or more, depending upon the flight altitude. Major highways tend to stand out at night because of the presence of numerous automobile headlights. As major highways stretch into the distance, the visual impression is of a string of randomly connected beads laid across the dark terrain. Less well-traveled roads are usually not easily seen at night unless the moonlight is bright enough to reveal them.

On clear *moonlit* nights the outlines of the terrain and other surface features are dimly visible; however, on extremely dark nights, terrain features are all but invisible except in brightly lighted populated areas. A pilot can often discern the outline of bodies of water on moonlit nights from the reflection of the moonlight from the surface.

LOCATING OTHER AIRCRAFT AT NIGHT

The position of other aircraft at night can be determined by scanning for position lights and anticollision lights. In fact, the presence of other aircraft may be easier to determine than during the day. However, the determination of the *heading* of the other aircraft and whether it is heading toward or away is somewhat more difficult.

Certain "rules of thumb" and relationships may be of assistance. Since the arrangement of the red and green aircraft position lights is the same as those on boats and ships, the old "red right-returning" axiom of sailing days is applicable. In other words, if the pilot observes a red and a green navigation light and the position of the lights is such that the red light is on the right, the aircraft is approaching as illustrated by airplane A in figure 8-9. Another helpful

hint is if the white position light on the tail can be seen, the aircraft is on a heading that will take it *away* from the pilot's immediate area. (See airplane B in Fig. 8-9.)

OVERFLIGHT OF THE AIRPORT

The general perspective of the airport at night and the location of runway boundary lights, threshold lights, taxiway lights, ramps, hangars, etc. can easily be determined by flying *directly over the airport.* This overflight should be made *at least 500 feet above* the existing traffic pattern altitude at uncontrolled airports and *at more than 3,000 feet AGL* at airports with operating control towers. This type of operation is valuable not

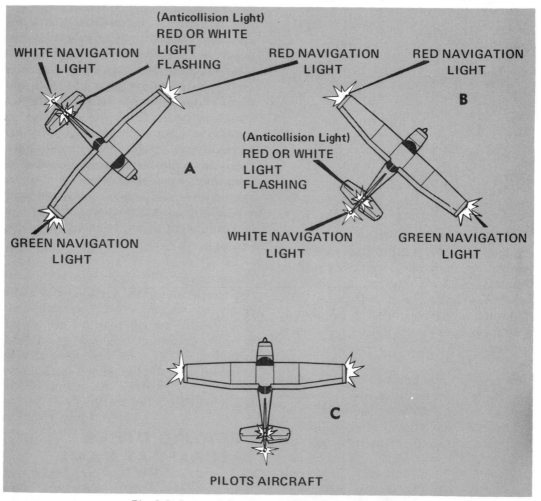

Fig. 8-9. Determining Direction From Position Lights

only at the home airport, but also during the time the pilot is building night proficiency and experience in approaching an unfamiliar airport.

AIRWORK

Most pilots agree that night flying is a *semi-instrument condition*. There are times when the discipline of an instrument-rated pilot is needed because the senses may urge the pilot to believe his "seat of the pants" sensations when his instrument readings are definitely telling him otherwise (and the *truth*).

Preparation for night flying should include a review of basic instrument flight techniques. This "under the hood" session should include straight-and-level flight, turns, climbs, climbing turns, descents, and descending turns. Many instructors require the student to make these turns to specific headings while under the hood in order to teach the pilot to respond to possible radar assistance. The "time honored" 180° turn is also excellent preparation for the possibility of flying into unanticipated cloud formations.

Often the instructor will ask a student to close his eyes and lower his head while the instructor puts the aircraft into *critical attitudes* and determines that the student can recover to straight-and-level flight unassisted by outside visual references. Therefore, an important night flying technique is *to rely on the instruments if disorientation should take place.*

WEATHER CONSIDERATIONS

The pilot operating at night must be especially attentive to signs of changing weather conditions. Pilots who are accustomed to daytime flight operations are generally not aware that it is extremely easy to fly into an overcast or cloud deck at night simply because such weather phenomena cannot easily be detected by direct visual observation.

There are several clues and suggestions that will assist the night pilot if he inadvertently flies into heavy haze, patches of clouds, or an overcast.

A pilot approaching an overcast can sometimes detect the presence of the overcast because it seems as though an *invisible black curtain* is slowly being pulled down ahead of the airplane and the lights in the distance wink out and disappear. Furthermore, a luminous glow or halo around the position lights indicates imminent or actual penetration of IFR weather conditions.

Further indirect visual clues to the presence of IFR conditions may be obtained by turning on the landing light for a short period of time. On a very clear night, the beam of a landing light is barely scattered by particles in the air; however, if there is considerable haze or the temperature and dewpoint are rapidly converging and cloud formation is imminent, some scattering of the beam will be noticed. Of course, if actual penetration of a cloud layer occurs, the landing light beam will be dispersed in all directions. If inadvertent penetration of IFR conditions does occur, the pilot should *calmly but immediately* initiate a 180° standard-rate turn in order to fly out of the weather condition.

One of the best insurance measures to permit true enjoyment of night flying is to obtain a thorough weather briefing prior to departure. Special attention should be paid to any indications in the weather briefing that would suggest the formation of clouds, overcast, fog, etc. while enroute or, for that matter, even while flying in the local area.

ENROUTE PROCEDURES

In order to provide improved margins of safety, the choice of high cruising altitudes is recommended. There are several reasons for this recommendation. *First*, range is greater at higher altitudes. *Second*, gliding distance is greater. *Third*, pilotage and radio navigation is often less difficult.

A major consideration in planning a night flight is to insure that an adequate supply of fuel is on board. A useful rule of thumb is to reduce the daytime range of the airplane by *one-third* when flying at night. Several advantages are gained from this suggestion. *First*, in cross-country flying, the pilot is not tempted to *stretch* his range. Pilots are sometimes led into this trap when they fail to check and determine that refueling facilities will be *open* at their estimated time of arrival, then find fuel unavailable and say to themselves, "Well, I can make it anyway!" *Second*, the additional fuel may be useful in avoiding or circumnavigating adverse weather.

The use of a subdued white cockpit light for reading charts is recommended since considerable information on charts is printed in red and disappears under the red cockpit lighting used in some aircraft. If a map reading light is not available in the airplane, a small flashlight should be carried for reading the charts. Special emphasis needs to be placed upon careful terrain clearance interpretation from the charts in order to insure adequate obstruction clearance when operating at night.

EMERGENCY LANDINGS

Stated simply and bluntly, the probability of finding a suitable off-airport field for an emergency landing at night is more or less a *matter of luck*. Even with bright moonlight, illumination is not sufficient to reveal the nature of the surface and terrain.

If a night forced landing is encountered, the same procedures as recommended for daytime would be adhered to and the landing light, if available, should be turned on during final approach to assist in avoiding obstacles in the final approach path.

Some pilots use a different route at night than in the daytime -- one that keeps them within reach of an airport as much of the time as possible. For example, *a course comprised of a series of 25⁰ zig-zags is only 10 percent longer than a straight-line course.* Assuming most light airplanes provide a glide ratio of 8:1, or better, when flying at 10,000 feet above ground level (AGL), it is possible to glide 16 miles and perhaps be within range of some airport most of the time.

Highways look like tempting emergency landing strips at night; however, only the four-lane type superhighway offers even reasonable assurance of the absence of powerlines alongside or across the highway. If a superhighway is selected, the landing should be made in the same direction as the flow of traffic. Since powerlines are invisible during a night approach to a highway, two-lane highways are almost automatically ruled out as possible emergency landing sites.

LANDINGS

In some respects, night landings are actually easier than daytime landings since the air is generally smooth and usually the disrupting effects of turbulence and excessive crosswinds are absent. There are a few special considerations and techniques that should be examined and perfected in order to properly qualify the pilot for landings after dark.

Certain carefully controlled studies have revealed that pilots have a tendency to *make lower approaches* at night than during the daytime. Therefore, careful consideration should be given to traffic pattern procedures and to the factors that will enable the pilot to maintain the proper glide slope angle when on final approach.

First of all, flying a carefully standardized approach pattern is recommended, using the altimeter and vertical speed indicator to monitor the rate of descent. If the downwind leg is flown at 800 feet, a square descent pattern beginning at the point *opposite* the touchdown point is recommended. When the intended touchdown point is "off the wing," power can be reduced to establish a descent rate of approximately 500 feet per minute; when the touchdown point is about 45° behind the pilot, he should turn base and be at an altitude of approximately 500 feet AGL.

After the base leg is flown and the aircraft "rolled out on final," the altitude should be around 200 feet AGL. The pattern should be flown at the same distance from the runway as during daylight to make possible a rectangular approach with a steady rate of descent. Maintaining the proper pitch attitude in order to stabilize the airspeed is as appropriate for night approaches as for those made during the daytime.

Many experienced pilots prefer to make power approaches when landing at night.

The runway lights provide an effective *peripheral* vision clue to gauge the approach and leveling-off phases of the landing. The runway lights, as seen in the peripheral vision, seem to rise and spread laterally as the pilot nears the proper touchdown point. (See Fig. 8-10.)

Naturally, most pilots use the landing lights for night landing; however, there is an inherent pitfall to be avoided. That is, the portion of the runway illuminated by the landing lights seems *higher* than the "black hole" surrounding it, tending to cause the pilot to level off high with an unduly hard landing as a possible result. Furthermore, to focus one's attention on the area *immediately in front* of the airplane is poor practice in any landing, but the arrangement of most landing lights tends to encourage this practice. Therefore, when using the landing lights, the pilot's sighting point should be at the forward limit of (and possibly beyond) the lighted area.

Proper preparation for night flight should include landings made both *with and without* the aid of the landing lights. The clues for the proper maintenance of the approach profile are derived from

Fig. 8-10. Night Landing

the altimeter indications, as previously discussed, and when on "final" from the perspective created by the size, shape, and patterns of the runway lights. Again, the altimeter and the vertical speed indicator should be checked against the position in the pattern to monitor "how the approach is going."

A "no light" or "blackout" landing may be made in the same manner as described or modified slightly in the flareout and touchdown phase. A normal approach is held until over the threshold where the flareout would normally begin; at this point, the airplane is slowed by using a slightly higher than normal nose position. With his hand on the throttle, the pilot adds power as needed and by maintaining a very shallow sink rate, literally "walks" the plane to the runway, flying a bit by the "seat of his pants." He should definitely guard against an increasing sink rate or the approach of a stall. If needed, the throttle is definitely, but measuredly, advanced since touchdown is imminent.

Proficiency in performing landings without landing lights requires practice. After the first few shakey approaches and touchdowns, they can even be fun. Certainly, familiarity with the technique is comforting should the landing light decide to "quit and go home" during some dark night approach.

ILLUSIONS

There are several illusions to which the pilot may be subjected during night approaches and landings. One is that the runway lights appear to form a *flat plane that can be mistaken for the runway* itself, as shown in figure 8-11. The runway is actually lower than the flat plane created by these lights. This is because the runway lights are usually mounted on short poles above the actual runway surface.

A common illusion that may be experienced while on the runway is that of moving forward faster than the actual

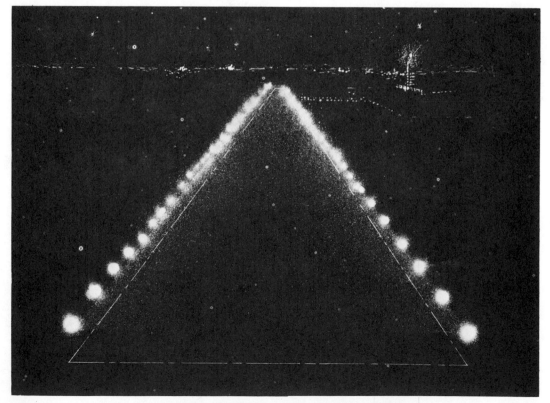

Fig. 8-11. Illusion Caused By Runway Lighting

speed. This illusion results from the fact that only *nearby* objects (boundary lights, centerline lights, etc.) can be seen clearly and they seem to move past the airplane quite rapidly.

Another illusion is that a *normal approach looks steeper at night*, creating an illusion of overshooting, which may actually result in undershooting and landing short. This illusion is thought to be responsible for the general tendency of pilots to make lower approaches at night, even though they believe they are on the correct approach profile.

The lack of clearly defined ground reference points in the approach path may induce pilots to maintain a higher than normal approach speed. Again, maintaining the correct approach pitch attitude which controls and stabilizes the airspeed is a principle that should be applied at night as judiciously as during the day.

Another illusion may be created by the *slope of the runway*. For example, a runway that slopes up and away from the threshold can make the pilot believe that he is too high. On the other hand, a pilot approaching the high end of a downward sloping runway may receive the illusion that the approach is too low. The remedies for these illusions are careful attention to flying a normal traffic pattern and using aids such as VASI lights where they are available.

VASI LIGHTS

Most visual approach indicators are designed with a final approach profile or

glide slope angle of about *2-1/2° to 4° above the ground.* The *visual approach slope indicator,* commonly termed VASI, is a system of visual lighting which permits the pilot to set up a stabilized descent on the proper glide slope while several miles away from the runway. Then, he receives continuous visual information that allows him to follow the correct approach path all the way to touchdown.

The visual approach slope indicator system is especially valuable to a pilot operating at night since most of the normal terrain profile clues used to determine the height of the aircraft on final approach are subdued or entirely lacking.

A short review of the design of the VASI system may aid the prospective night pilot. A common VASI installation consists of 12 identical light boxes arranged with two identical sets on *each side* of the runway. The first set, or row, of VASI lights is called the *downwind* row and is placed about 600 feet beyond the threshold. The second row is called the *upwind* row and is located about 1,300 feet from the threshold. (See Fig. 8-12.) The four groups of lights tend to form a rectangle around the touchdown zone and provide a visual aiming point toward the middle of this rectangle.

Each VASI box has a filter that effectively splits the light beams into a white segment above a certain angle and a red segment below this angle. If an aircraft on final approach *overshoots* a VASI box, the light beam appears *white* to the

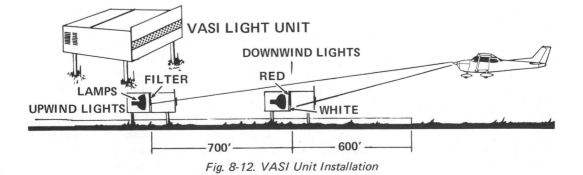

Fig. 8-12. VASI Unit Installation

pilot. On the other hand, an *undershoot* reveals the *red* segment of the light beam.

Since the objective is to land *between* the upwind row and the downwind row, the upwind row of lights should appear red and the downwind row of lights should look white when viewed by the pilot on the proper final approach profile. *Red over white*, then, is the proper light combination observed when the aircraft is on the glide path. (See Fig. 8-13.)

If both rows appear white, the aircraft is too high and will overshoot the intended landing area. (See Fig. 8-14.) If both rows look red, the aircraft is too low, will undershoot, and the landing will be short, as shown in figure 8-15. When the approach path of the aircraft is not stabilized and the aircraft is passing *through* the glide slope, the pilot can see the lights change color. For example, if the aircraft is above the glide slope and all the lights appear white, the descent rate should be increased. As the aircraft approaches the proper glide slope, the upwind row of lights will *transition* from white to pink and then to red. Thus, after the aircraft is once again on the proper glide slope, the upwind row of lights will be red and the downwind row will appear white.

Conversely, if the aircraft descends below the glide slope, the downwind row will transition from white through pink to red to provide an "all red" indication. Also, if the aircraft is at pattern altitude while still 10 to 15 miles from the threshold, all of the VASI lights appear red since the VASI glide slope angle at that point passes above the aircraft. By flying toward the runway at a constant pattern altitude, the pilot can see the downwind row gradually change from red to pink to white and he can transition from level flight to final-approach descent as the glide slope is intercepted.

Where VASI lights are available, they should be used because they make a precise approach much easier during night

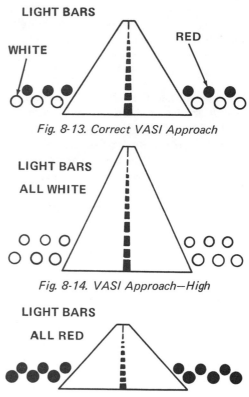

Fig. 8-13. Correct VASI Approach

Fig. 8-14. VASI Approach—High

Fig. 8-15. VASI Approach—Low

operations or when the visibility is restricted. They also provide a visual descent profile that *assures* proper obstruction clearance in the approach area. VASI installations also provide visual reference for approaches made over water or other featureless terrain where adequate visual references are not available or perhaps misleading.

ACCEPTABLE PERFORMANCE

The student will be evaluated on the significance of items peculiar to night flight, proper use of power during approach, use of landing lights, safe climb and approach paths, and safe taxi speeds. Also, the recognition of position with respect to other aircraft, dangers of vertigo, vigilance for aircraft on the ground and in the air, and wake turbulence avoidance will be evaluated. The student should be able to demonstrate night VFR navigation, as required for cross-country flight operations.

SECTION B—CROSS-COUNTRY OPERATIONS

Today's light aircraft offers the pilot and his passengers the greatest mobility known to modern man. In order to utilize this mobility safely, a pilot must know and understand the operating principles concerning cross-country flight. With adequate thought and preflight action, the pilot can complete a flight with confidence, ease, and safety.

PLANNING

When preparing for a cross-country flight, a pilot is expected to have sectional charts for the intended route plus an alternate route, a plotter, computer, flight planning sheets, and access to Parts 1, 2, 3, and 4 of the *Airman's Information Manual* (AIM). In addition, the pilot should also have a basic knowledge relating to sectional charts, meteorology, radio technique, airplane performance, and the flight computer.

Federal Aviation Regulation Part 91.5 states that each pilot in command shall, before beginning a flight, familiarize himself with all available information concerning that flight. This information must include, for a flight not in the vicinity of an airport, available weather reports and forecasts, fuel, runway requirements, pilot qualifications, and alternatives available if the planned flight cannot be completed.

While filing a flight plan is not required by the FARs, it is dictated by good operating practice since it aids in search and rescue. As preparations are made, the necessary materials should be gathered and assembled at the planning work area prior to outlining the flight. Organization and planning prior to flight as well as in actual flights may eliminate embarrassing or disasterous situations later.

Within the privileges of Part 61, certain limitations are placed upon the pilot in command. Included are the requirements for recency of flight experience and pilot certification. A review of the Federal Aviation Regulations (FARs 61, 91, and 430) is recommended.

ROUTE SELECTION

Factors that should be considered when selecting the route of flight are: aircraft capability, location of radio navigational facilities, associated weather conditions and, if applicable, height of enroute terrain obstructions. When plotting the course, a straight line should be drawn from the center of the departure airport to the center of the first point where a direction change occurs.

When available, the VOR is utilized to aid in the navigation process. A pilot may select a radial which most closely aligns with his course (See Fig. 8-16) to help keep him oriented on his flight planned route; or, he may also choose

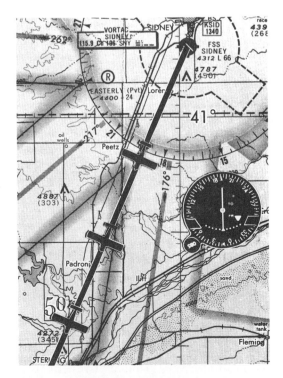

Fig. 8-16. Tracking VOR Course

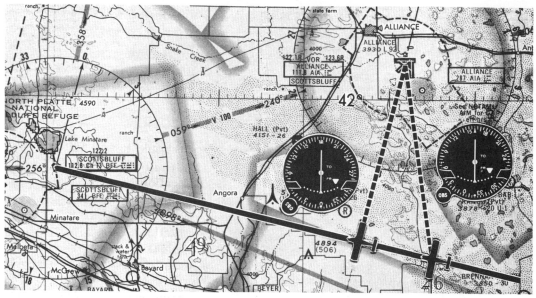

Fig. 8-17. Using VOR Radials As Checkpoints

several VOR radial intersections along his true course to substitute for visual checkpoints when clearly defined landmarks are not available. (See Fig. 8-17.)

The pilot should examine his route closely to insure compliance with airspace restrictions. As a supplement to the sectional chart, the *Airman's Information Manual* (AIM) should be consulted for special operations such as oil burner routes, wake turbulence, bird hazards, etc., and for additional airport information.

Special efforts should be made to avoid rough terrain or areas having rapid changes in elevations during high wind conditions. When a body of air flows around such obstructions, irregular whirls or eddies form and create turbulence. (See Fig. 8-18.) The severity of the abrupt jolts or bumps depends upon the wind velocity, aircraft weight, wing loading, airspeed, and attitude. Turbulence criteria is expressed as either being light, moderate, severe, or extreme.

Light - a condition where unsecured objects may be displaced slightly.

Moderate - very similar to light turbulence but of greater magnitude.

Severe - a condition creating large, abrupt jolts with large changes in altitude and/or attitude and wide fluctuations in indicated airspeed.

Extreme - a condition where the aircraft is tossed about violently, making control impossible.

Turbulence, or any other areas of unusual weather, should be reported by the pilot to the nearest flight service station (FSS) or air route traffic control center (ARTCC), giving

> *location(s),*
>
> *time (GMT),*
>
> *intensity, and*
>
> *proximity to clouds.*

Fig. 8-18. Wind Effects Over Mountains

By using information from aircraft in flight, special reports can be compiled to form data known as pilot reports (PIREPs). This information informs other pilots flying along or near the same routes of impending flight conditions. The *Airman's Information Manual* (AIM) should be consulted for further information concerning PIREPs.

The immediate area surrounding the pilot's terminal airport should be examined for any obstructions by referring to Part 3 of the AIM. This publication provides the pilot with all the necessary data for safely planning his flight. The sectional charts should be reviewed to determine areas of controlled, restricted, or prohibited airspace enroute along with its respective controlling agency.

Terrain elevation along the route of flight should be determined to assure adherence to the proper VFR altitudes spelled out in the AIM. The VFR hemispheric cruising rules state that all aircraft operating above 3,000 feet AGL shall fly at odd-thousands plus 500 feet when traveling on a magnetic course of 0° through 179°, and even-thousands plus 500 feet when flying magnetic courses between 180° and 360°.

Special emphasis should be placed on aircraft performance, as the optimum cruising altitude for single-engine aircraft equipped with a fixed-pitch propeller is generally 7,500 feet MSL, while optimum cruise for aircraft having a constant-speed propeller is 9,000 feet MSL.

WEIGHT AND BALANCE

When reviewing the cross-country flight, the pilot must insure that the aircraft is within its legal weight and balance limits since misloading of fuel, baggage, or passengers can severely hamper the aircraft's overall performance. Almost every aspect of weight and balance is influenced by the weight and contents of the aircraft. Loading an aircraft too heavily

will decrease performance to a dangerous level by increasing the risk of structural damage caused by turbulence, hard landings, or metal fatigue.

It should be stressed that the pilot not only consider weight, but also load distribution. If passengers, cargo, or fuel are not all in their respective positions and within the CG limits, the airplane will be off balance, reducing stability and control response and general deterioration of aircraft performance.

A procedure to compute weight and balance can be determined by referring to the aircraft owner's manual. The listing of optional equipment such as radios, extra fuel tanks, etc., is included in the equipment list. Also provided is the weight of each item and its location with respect to the specified reference (datum) point on the airplane. (See Fig. 8-19.)

WEATHER BRIEFING

The pilot will recall that weather information can be obtained from the flight service station or the National Weather Service forecast office by various means of communication. The list includes two-way radio or telephone. A personal visit by the pilot is suggested in order to provide a more total picture of the current existing weather conditions. In addition to weather maps and charts on display, the weather station will also provide the pilot with area forecasts, terminal forecasts, hourly reports, winds aloft reports, and personal assistance. (See Fig. 8-20.)

If a personal visit is impractical, a briefing can be received by a telephone call to the flight service station or National Weather Service forecast office. These telephone numbers are listed in Part 2 of the *Airman's Information Manual* (AIM). (See Fig. 8-21.) When requesting a telephone briefing, the pilot should identify himself as a pilot and indicate to the briefer the proposed route, destination, estimated time of departure (ETD), pilot

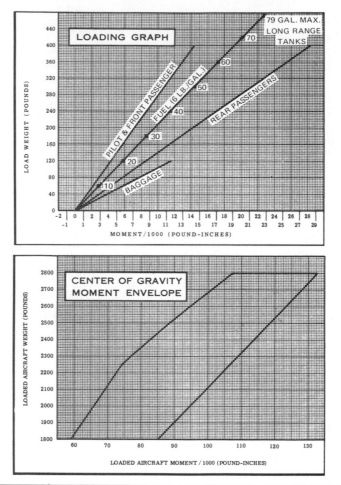

SAMPLE LOADING PROBLEM	Sample Airplane			Your Airplane	
	Weight (lbs)	Moment (lb - ins. /1000)		Weight	Moment
1. Licensed Empty Weight (Sample Airplane) ...	1625.1	55.5		1707.5	61.0
2. Oil - 12 Qts.*	22.5	-0.3		22.5	-0.3
3. Pilot & Front Passenger	340.0	12.2		185.0	6.6
4. Fuel- (60.0 Gal at 6#/Gal)	360.0	17.3		280.0	13.5
5. Rear Passengers	340.0	24.1		330.0	23.4
6. Baggage	112.4	10.9		100.0	9.6
7. Total Aircraft Weight (Loaded)	2800.0	119.7		2625.0	113.8

8. Locate this point (2800 at 119.7 on the center of gravity envelope, and since this point falls within the envelope the loading is acceptable.

*Note: Normally full oil may be assumed for all flights.

Fig. 8-19. Typical Aircraft Weight And Balance Computation Forms

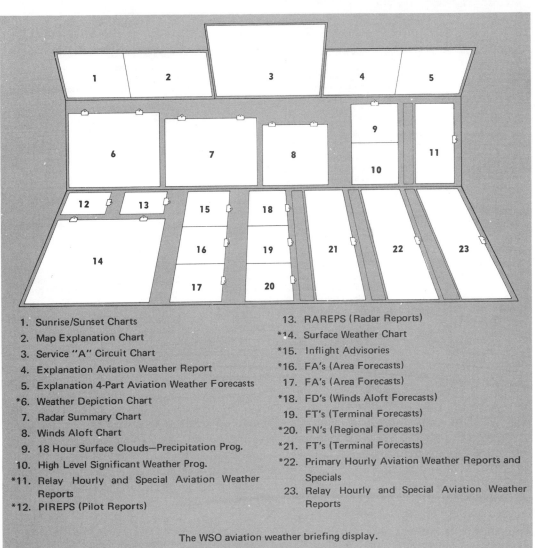

1. Sunrise/Sunset Charts
2. Map Explanation Chart
3. Service "A" Circuit Chart
4. Explanation Aviation Weather Report
5. Explanation 4-Part Aviation Weather Forecasts
*6. Weather Depiction Chart
7. Radar Summary Chart
8. Winds Aloft Chart
9. 18 Hour Surface Clouds—Precipitation Prog.
10. High Level Significant Weather Prog.
*11. Relay Hourly and Special Aviation Weather Reports
*12. PIREPS (Pilot Reports)

13. RAREPS (Radar Reports)
*14. Surface Weather Chart
*15. Inflight Advisories
*16. FA's (Area Forecasts)
17. FA's (Area Forecasts)
*18. FD's (Winds Aloft Forecasts)
19. FT's (Terminal Forecasts)
*20. FN's (Regional Forecasts)
*21. FT's (Terminal Forecasts)
*22. Primary Hourly Aviation Weather Reports and Specials
23. Relay Hourly and Special Aviation Weather Reports

The WSO aviation weather briefing display.

*of special interest to the private pilot

Fig. 8-20. Typical National Weather Service Briefing Display

qualifications, aircraft type, and whether the flight will be conducted VFR or IFR. This information will help the National Weather Service personnel in providing the pilot with more precise and accurate information along the intended route.

As an additional means of obtaining a weather briefing, a transcribed weather broadcast (TWEBs) is provided. This equipment is furnished at selected FAA flight service station sites. Pertinent meteorological and Notice to Airmen data is recorded on tapes and broadcast continually over the low frequency band 200 to 415 kHz, while live broadcasts are made over the VOR bands 112.0 to 117.9 MHz 15 minutes past each hour. (See Fig. 8-22.)

The continuous recorded broadcasts are made from a series of individual tape recordings. The first three tapes identify the station, giving general weather forecast conditions in the area, pilot reports (PIREPs), and radar reports, when available. The remaining tapes contain weather at selected locations within a 400 mile radius of the reporting point. Changes, as they occur, are transcribed onto the tape. The scheduled weather broadcasts,

FSS—CS/T AND WEATHER SERVICE TELEPHONE NUMBERS

Flight Service Stations (FSS) and Combined Station/Tower (CS/T) provide information on airport conditions, radio aids and other facilities, and process flight plans. CS/T personnel are not certificated pilot weather briefers; however, they provide factual data from weather reports and forecasts. Airport Advisory Service is provided at the pilot's request on 123.6 by FSSs located at airports where there are not control towers in operation. (See Part 1 ARRIVALS.)

The telephone area code number is shown in parentheses. Each number given is the preferred telephone number to obtain flight weather information. Automatic answering devices are sometimes used on listed lines to given general local weather information during peak workloads. To avoid getting the recorded general weather announcement, use the selected telephone number listed.

★ Indicates Pilot's Automatic Telephone Weather Answering Service (PATWAS) or telephone connected to the Transcribed Weather Broadcast (TWEB) providing transcribed aviation weather information.

◆ Indicates a restricted number, use for aviation weather information

■ Call FSS for "one call" FSS/WBO briefing service.

⌗ Automatic Aviation Weather Service (AAWS).

Location and Identifier		Area Code	Telephone
ALABAMA			
Anniston ANB	FSS	(205)	831–2303
Birmingham BHM	FSS	(205)	595–6151 ■
	FSS	(205)	595–2101 ★
Dothan DHN	FSS	(205)	794–6683
Huntsville HSV	WB	(205)	772–9308 ◆
Mobile MOB (Bates)	FSS	(205)	344–3610
	WB	(205)	342–2762 ◆
Montgomery MGM (Dannelly)	FSS	(205)	269–4368
	WB	(205)	265–0589 ◆
Muscle Shoals MSL	FSS	(205)	383–6541 ■
	FSS	(205)	381–2500 ★

Location and Identifier		Area Code	Telephone
ARKANSAS			
El Dorado ELD (Goodwin)	FSS	(501)	863–5128
Fayetteville FYV (Drake)	FSS	(501)	HI 2–8277
Ft. Smith FSM	CS/T	(501)	MI 6–7868/69
	WB	(501)	646–5731
Harrison HRO	FSS	(501)	EM 5–3433
Jonesboro JBR	FSS	(501)	WE 5–3471
Little Rock	WB	(501)	374–1546 ◆
Pine Bluff PBF (Grider)	FSS	(501)	JE 5–0652
Texarkana TXK	FSS	(501)	774–4151 ■

Fig. 8-21. Excerpt Of Airman's Information Manual

over the VOR system, are compiled from reporting stations within approximately 150 miles from the broadcast station.

SELF-HELP FACILITIES

Many remote locations have teletypewriter connections available to pilots on a "serve yourself" basis. In this situation, it is to the pilot's advantage to be familiar with the interpretation of codes, contractions, and abbreviations used on the teletypewriter units.

WEATHER ANALYSIS

During the planning stage of a flight, the pilot is encouraged to take advantage of the help available to him at the National Weather Service forecast office and flight service station (FSS). A forecaster will thoroughly brief the pilot on any information necessary or desired. By using the surface weather maps, aviation reports, area forecasts, and winds aloft reports, the pilot can get a general con-

cept of weather that may be encountered during his flight. Even though the pilot often interprets the weather from the charts and data supplied, the professional forecaster should be used if available.

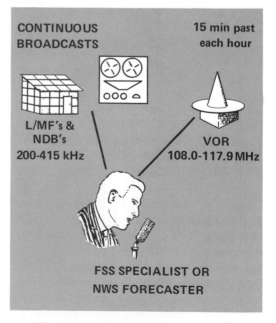

Fig. 8-22. Aviation Weather Broadcasts

PILOT'S WEATHER CHECKLIST

The careful pilot will prepare a checklist for requesting weather information. This is not only good operating practice, but also reassurance that nothing is left out. Items that should be checked are listed below.

1. Preflight Weather—A knowledge and understanding of existing conditions prior to takeoff is of prime importance and should include winds, ceiling, temperature, visibility, type and extent of cloud cover, and runway conditions. As a convenience to the pilot, a weather checklist is provided on the back of each FAA flight plan form.

2. Enroute Weather — A pilot should know and understand types and trends of fronts lying along the route, their location, intensity, direction, speed, cloud types, and extent of cover. He should also know the locations of severe weather areas, such as thunderstorms, hail, icing, turbulence, and restrictions to visibility. The selected route should be compatible with the pilot's knowledge and experience level.

3. Destination And Alternate Weather — This item includes visibility, type, and extent of cloud cover, precipitation (type and intensity), forecast temperature/dewpoint spread, hazards that may be encountered (thunderstorm, icing, etc.), surface winds (direction, velocity, and tendencies), winds aloft forecast, humidity, turbulence, runway conditions, and forecast conditions for the arrival airport.

Above all, the pilot must view his own experience level based on sound judgment, not compulsion. If there is any doubt as to the safety of the flight, it should not be made.

METHODS OF AIR NAVIGATION

Navigation is simply following a preselected route from one place to another. Traditionally, cross-country air navigation is divided into three categories: *pilotage, dead reckoning,* and *radio navigation.*

PILOTAGE

Pilotage is a method of visual navigation. It is accomplished by reference to land features and cultural symbols found on the legend panel of aeronautical sectional charts. A course line is drawn on the chart between two points and is followed by using landmarks for checkpoints. Checkpoints are used to direct the pilot to his destination, to make heading corrections, and to adjust for the effects of wind drift.

DEAD RECKONING

Dead reckoning is a system of mathematical navigation. It is the technique of determining the aircraft's position over a fixed point on the earth's surface by performing calculations based on time, speed, direction, and distance. Pilotage is used in conjunction with dead reckoning to determine that values assumed when initially planning the flight are correct and that the aircraft is following the selected course.

RADIO NAVIGATION

Radio navigation is a supplement to dead reckoning and pilotage navigation and consists of following a course formed by radio signals. These signals are transmitted through a VHF (very high frequency) radio facility (VOR) to the aircraft's navigation receiver.

The wise cross-country pilot will plan his flight so as to utilize all three methods. A combination of pilotage, dead reckoning, and radio navigation should be used whenever practical to reduce the possibility of error.

CHECKPOINTS

Prior to flight, the pilot should be familiar with the required sectional charts. When selecting checkpoints, the pilot should be certain the selected features are depicted on the chart and can be readily identified from the air. Figure 8-23 describes how to select checkpoints.

After time and distance computations are made during flight, the pilot should have an estimated time of arrival (ETA) to his next checkpoint. By comparing time and visual observation of the checkpoint, it is easy to ascertain the exact position of the aircraft relative to the true course line. Checkpoints should be selected that are close to the flight path course and at convenient intervals approximately 15 to 30 miles apart, depending upon the number of checkpoints available and aircraft speed.

FLYING THE TRIP

At the completion of the preflight planning, the pilot should have in mind and on paper a definite plan of immediate and alternative action for the trip. Existing and forecast weather should have been checked, as well as tentative fuel stops, weight and balance, and any changes in loading made to insure that the aircraft is within the center of gravity (CG) envelope. A flight plan should be filled out and filed on the ground - either in person or by phone— to avoid radio congestion while airborne. The pilot should have a mental plan for departing the airport traffic area and becoming established on course.

DEPARTURE

After departing from an airport with a UNICOM or FSS, a pilot should monitor the frequency used for takeoff until clear of the airport area; or, if operating in controlled airspace, until frequency change approval from the control tower is received. At uncontrolled airports equipped with UNICOM, 122.8 MHz should be monitored, while 122.9 MHz is used primarily at airports not utilizing UNICOM.

The departure phase of the trip is one of the busiest portions of the flight. The pilot is maneuvering the aircraft, establishing his departure heading, opening the flight plan, and maintaining a constant vigil for avoidance of other aircraft. This is when thorough flight planning "pays off," since the majority of the paperwork has already been accomplished allowing the pilot to concentrate his attention outside the cockpit.

During the initial climb, the pilot should bear in mind that as altitude is gained, air becomes less dense (thinner), requiring leaning of the fuel mixture. When this is accomplished, maximum available power is attained along with best economy.

During the climb to cruise altitude, a blind spot is created directly ahead of the aircraft due to the nose-high attitude. This hazard may be alleviated somewhat by using the normal (cruise) climb speed specified in the aircraft flight manual. Since this speed is generally higher than the best angle- or rate-of-climb speed, the pitch attitude is lower, resulting in improved visibility over the nose.

To clear the area directly ahead of the aircraft, the pilot should initiate shallow-banked turns right and left while climbing. This procedure allows the pilot to scan the area normally blocked by the nose-high attitude of the aircraft. Throughout the entire flight, the pilot should request assistance from his passengers in locating other aerial traffic. While adding safety to the flight, passenger interest increases, reducing the possibility of boredom.

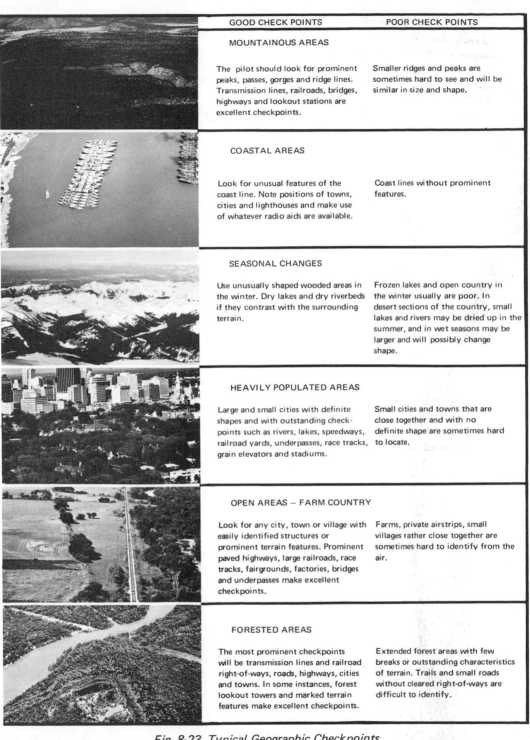

	GOOD CHECK POINTS	POOR CHECK POINTS
	MOUNTAINOUS AREAS	
	The pilot should look for prominent peaks, passes, gorges and ridge lines. Transmission lines, railroads, bridges, highways and lookout stations are excellent checkpoints.	Smaller ridges and peaks are sometimes hard to see and will be similar in size and shape.
	COASTAL AREAS	
	Look for unusual features of the coast line. Note positions of towns, cities and lighthouses and make use of whatever radio aids are available.	Coast lines without prominent features.
	SEASONAL CHANGES	
	Use unusually shaped wooded areas in the winter. Dry lakes and dry riverbeds if they contrast with the surrounding terrain.	Frozen lakes and open country in the winter usually are poor. In desert sections of the country, small lakes and rivers may be dried up in the summer, and in wet seasons may be larger and will possibly change shape.
	HEAVILY POPULATED AREAS	
	Large and small cities with definite shapes and with outstanding check-points such as rivers, lakes, speedways, railroad yards, underpasses, race tracks, grain elevators and stadiums.	Small cities and towns that are close together and with no definite shape are sometimes hard to locate.
	OPEN AREAS — FARM COUNTRY	
	Look for any city, town or village with easily identified structures or prominent terrain features. Prominent paved highways, large railroads, race tracks, fairgrounds, factories, bridges and underpasses make excellent checkpoints.	Farms, private airstrips, small villages rather close together are sometimes hard to identify from the air.
	FORESTED AREAS	
	The most prominent checkpoints will be transmission lines and railroad right-of-ways, roads, highways, cities and towns. In some instances, forest lookout towers and marked terrain features make excellent checkpoints.	Extended forest areas with few breaks or outstanding characteristics of terrain. Trails and small roads without cleared right-of-ways are difficult to identify.

Fig. 8-23. Typical Geographic Checkpoints

ENROUTE

The enroute portion begins when the aircraft arrives at the preselected cruising altitude. At this point, the aircraft is leveled off and the power is set to a cruise power (with best mixture setting) to obtain a maximum true airspeed with low fuel consumption.

The pilot should be aware of his exact position at all times. The importance of the pilot knowing his precise location can be understood in the event of a forced landing. Should the remote possibility of a forced landing occur, the pilot has enough on his mind without trying to locate himself on the chart. If feasible, the pilot should broadcast his position to the nearest flight service station, tower or ARTCC.

If unexpected weather is encountered, the pilot must know his position in order to plan an alternate route around the existing weather. When circumnavigating a large weather system, it is easy for a pilot to become disoriented and lost if careful examination of checkpoints is not made.

When establishing his position, the pilot should use two or more references to establish positive identification. Roads, railroads, racetracks, mines, factories, rivers, lakes, and water towers are good examples of additional checkpoints. This pilotage method of navigation will become second nature after passing the first few checkpoints.

When flying cross country, it is not uncommon for a pilot to miss a checkpoint. This may occur for several reasons. The checkpoint selected may not have been a good one and hard to spot. The pilot may be off course or, on the other hand, he may have flown directly over the checkpoint; in which case, the airplane itself may have blocked the pilot's view of the checkpoint. The latter situation is remedied by the pilot flying slightly to the right of course so checkpoints are where the pilot can readily pick them out. In any event, the pilot should not spend a lot of time locating "lost checkpoints," but instead should continue flying the original heading which has kept the aircraft on course up to that point. If an accurate flight log has been kept, a close ETA to the destination airport can be made.

If the pilot becomes lost, contact should be established with the nearest flight service station or by transmitting on the international emergency frequency of 121.5 MHz.

Although position reports are not required for VFR flight plans, reporting periodically to FAA flight service stations along the route is good operating practice. Position reporting serves as a convenient time for the pilot to extend his flight plan should it become necessary.

When making a position report, certain information should be included and if presented in a logical sequence, avoids confusion. When contacting a flight service station, the following information should be provided in the position report:

1. aircraft identification,

2. position,

3. time over position,

4. type of flight plan (IFR or VFR), and

5. destination.

ENROUTE WEATHER INFORMATION

Direct pilot-to-weather-briefer service is available by radio contact with any FAA flight service station. Flight service specialists are qualified and certified by the National Oceanic Atmospheric Administration/National Weather Service (NOAA/NWS) as pilot weather briefers. They are not authorized to make original forecasts, but are permitted to adapt, translate, and interpret available forecasts and reports. They also assist the pilot in selecting an alternate route if unfavorable weather exists. It is not necessary to become a communications expert in order to state one's intentions. A brief call in the pilot's own words will receive immediate attention.

All flight service stations having voice facilities on radio ranges (VORs) or radio beacons (NDBs) broadcast weather reports and Notices to Airmen information 15 minutes past each hour from reporting points within approximately 150 miles of the broadcast station. SIGMETs and AIRMETs are issued by the NWS, advising the pilot of hazardous flying conditions.

AIRMETs are advisories issued to light aircraft when actual or anticipated deviations from the weather previously predicted in the latest terminal area forecasts prevail. SIGMETs provide information to all aircraft when severe weather conditions, such as thunderstorms, squall lines, mountain waves, etc. prevail.

OPERATIONS NEAR WEATHER

The pilot is able to obtain weather information from many sources while airborne and, therefore, avoid unfavorable weather conditions. In addition, he should be aware of the problems that may arise when circumnavigating weather.

THUNDERSTORMS

Turbulence, hail, rain, snow, lightning, up- and downdrafts, and icing conditions are all present in thunderstorms. Also, there is some evidence that the maximum turbulence exists at the core of a thunderstorm, although recent studies show little variation of turbulence intensity with altitude.

There is no useful correlation between the external visual appearance of a thunderstorm and its severity. The visible thunderstorm cloud is only a portion of the associated turbulence whose up- and downdrafts often extend far beyond the visible storm cloud. Severe turbulence can be expected up to 20 miles from the thunderstorm. This distance decreases in less severe storms.

Considerable turbulence may also be present beneath the thunderstorm. This is especially true when the relative humidity is low in any one layer between the surface and 15,000 feet. The lower altitudes are characterized by strong, outflowing winds with severe turbulence. The only safe rule for thunderstorm flying is "stay away." Give them a wide berth.

TURBULENCE

Since turbulence may be encountered, the pilot should know how to cope with it. When turbulence is expected, all unsecured objects should be tied down or placed in cargo nets. The pilot should be certain all seatbelts and shoulder harnesses are tightly buckled. The aircraft should be slowed to maneuvering speed, or the turbulent airspeed specified in the owner's manual, thus avoiding the possibility of overstressing the airplane. When encountering turbulence, the pilot should *not* attempt to hold altitude precisely. Since the vertical velocity of the updrafts and downdrafts are generally so intense that regardless of the aircraft's attitude or power setting, large changes in altitude will occur. Nonetheless, the pilot should attempt to maintain a level flight attitude with small power changes.

In turbulence, the airspeed indications will fluctuate rapidly causing unreliable readings. The pilot should attempt to maintain a level flight attitude and disregard airspeed indications.

OVER A CLOUD LAYER

When a pilot approaches an area of cloud cover, many factors must be taken into consideration before an alternate plan of action is taken. These factors include:

1. *height of the clouds above ground level,*
2. *thickness of the cloud layer,*
3. *how far it extends,*
4. *types of clouds (broken or overcast),*
5. *method of navigation being used,*
6. *ragged or smooth tops, and*

7. *whether the pilot will be able to descend to his destination while remaining in VFR conditions.*

If the cloud layer is high above the ground with an extensive broken or overcast layer, the pilot should fly at least 500 feet under it. If the cloud layer is low, the pilot should not attempt to fly over it unless he has determined that the destination weather is at least at VFR minimums and improving.

The pilot should be at least 1,000 feet above the clouds while operating on top of a broken or overcast layer. In this situation, the pilot will be dependent upon the VOR radio receiver as his primary means of navigation. If the top of the cloud layer is ragged or sloped, reference to the aircraft instruments will be necessary to maintain the desired flight attitude.

In the event of power failure "on top," the pilot should establish a normal glide to keep the rate of descent to a minimum while providing the maximum forward travel distance. Following this procedure, an effort should be made to locate the cause of engine failure and attempt to restart. Above all, the pilot should fight "panic" and maintain a normal wings-level glide, especially if descending through an overcast on instruments is the only alternative.

FUEL MANAGEMENT

The complexity of fuel management depends primarily upon the aircraft being operated. In most modern light aircraft, the fuel system is a simple arrangement incorporating a single OFF-ON or LEFT/ RIGHT/OFF selector valve and a fuel gauge for each fuel tank, as shown in figure 8-24. As aircraft performance increases, generally so does the complexity of the fuel system.

A complex system may include the use of auxiliary fuel tanks, fuel boost pumps, fuel flow meters, and certain additional

procedures for safe and efficient fuel management. A pilot should become thoroughly familiar with the fuel system and its operation prior to flight by reviewing the aircraft's owner's manual. He should be aware of the capacity of each tank and the rate at which the fuel is being consumed with a definite plan for balancing the fuel load during flight. In addition, any procedures used for emergency operation of the fuel system should be an automatic reaction developed by continued practice.

PASSENGER CONSIDERATION ENROUTE

As veteran pilots and passengers have found, cross-country flying may provide long periods of relative inactivity. To maintain interest, the pilot should perform regular inflight checks. This activity will keep the pilot alert and informed of his progress. The passengers may be asked to help locate checkpoints and other air traffic which will aid the pilot and add interest to the trip. One other method in combating fatigue and sustaining the interest of passengers is to view scenic points along the flight. Slight deviations in course may be desired to provide a more scenic route.

ARRIVAL

The arrival phase of flight begins at the point of "letdown" to the destination airport. Letdown should be planned so the aircraft reaches the airport traffic pattern altitude just prior to entering the pattern. A rate of descent should be selected that is comfortable to the passengers (approximately 300 to 500 feet per minute).

A common technique when descending is to lower the nose to maintain a shallow descent attitude while maintaining cruise power. The increase in airspeed offsets the slower airspeed used in the initial climb. It should be remembered, however, that while descending, the mixture should be richened periodically due

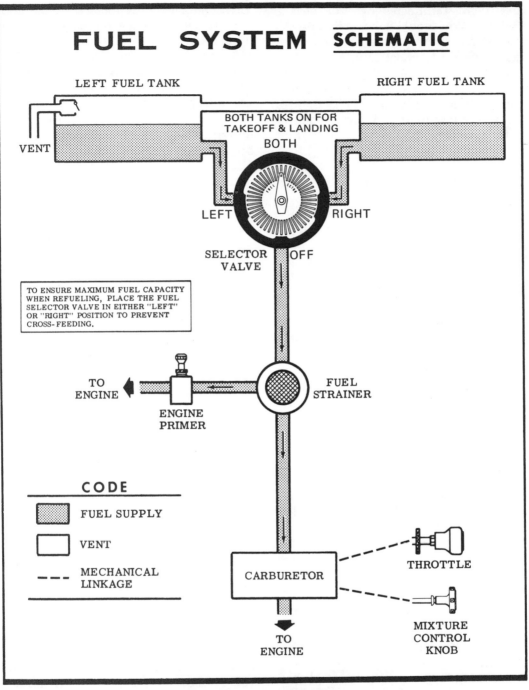

Fig. 8-24. Typical Fuel System

to the increased density of air at lower elevations.

When operating aircraft equipped with a constant-speed propeller, the manifold pressure must be reduced approximately one inch per 1,000 feet if the pilot is maintaining a constant power setting during descent. If an aircraft with a fixed-pitch propeller is flown, generally it is not equipped with a manifold pressure gauge; therefore, the limiting factor in this situation is engine r.p.m. Slight power (r.p.m.) reductions during descent are necessary in order to maintain a constant power setting.

If a flight terminates over or near a VOR, caution should be exercised due to converging aircraft generally found in and around a terminal area. FAA statistics show that about 50 percent of the near-miss incidents reported occurred over a VOR facility in clear skies with unrestricted visibility. A great majority of the aircraft were utilizing the VOR as a navigational aid.

CLOSING THE FLIGHT PLAN

After descent has been initiated, the pilot should contact the nearest flight service station and close his flight plan. When in contact with the facility, the pilot should "close out" the flight plan by first providing aircraft identification, followed by his position and his request. This procedure eliminates the possibility of the pilot forgetting to close his flight plan by telephone after landing. The pilot should keep in mind that when his flight plan is terminated, the benefits of immediate search and rescue efforts are forfeited.

CONTACTING THE DESTINATION AIRPORT

After communicating with "flight service," the pilot should contact the destination airport for landing instructions or advisories. For VFR aircraft operating at a controlled airport, advisory information may be provided by radar and non-radar approach control facilities. The controller issues wind direction and velocity, runway in use, traffic, and NOTAM information.

Initial radar contact is usually established within 30 statute miles of the facility. Radar contact points are based on time and distance rather than landmarks. Compliance with this procedure is not mandatory, but pilot participation is encouraged.

The pilot of an aircraft arriving at an airport where flight service, a tower, or

UNICOM facility is located should contact the facility when approaching within 15 miles of the airport to receive advisory information or landing instructions. This not only notifies personnel at the airport of the pilot's intentions, but also informs other traffic in the area as to the location of the arriving aircraft. An aircraft entering the traffic pattern at an airport serviced by a flight service station or UNICOM station should report when entering downwind, base leg, and on final approach. This is a good operating procedure as well as a good safety practice.

A pilot arriving at an airport that is not equipped with a tower, flight service station, or UNICOM may receive airport information from other aircraft operating on the airport over the frequency 122.9 MHz. If unable to establish contact or if loss of communication occurs, the pilot may fly over the airport (at least 500 feet above the pattern altitude) in order to establish the airport arrangement, wind direction, and runway in use. The pilot should then fly away from the airport, descend to the traffic pattern altitude, enter the pattern, and land. While flying the pattern, the pilot should broadcast his position and intentions "into the blind" on frequency 122.9 MHz for the benefit of other pilots in the airport area.

TERMINAL RADAR PROGRAMS FOR VFR AIRCRAFT

There are several programs which the VFR pilot may utilize when operating in a terminal radar environment. The use of these programs by VFR aircraft generally requires a pilot request. The fulfillment of this request depends on controller workload and communications equipment available. Pilots operating in accordance with visual flight rules, not equipped or trained for IFR flight, should be familiar with radar procedures when requesting assistance, since it is the responsibility of the pilot to remain in VFR conditions. In many cases, the radar controller is unable to determine if flight in instrument conditions will result from his instructions.

Stage I Service (Radar Advisory Service for VFR Aircraft)

Stage I facilities provide traffic information with limited vectoring to VFR aircraft. This service may be provided when requested by the pilot or with pilot concurrence when suggested by ATC. Pilots of arriving aircraft should contact approach control on the published frequency found in the AIM, giving their respective position, altitude, radar beacon code (if transponder equipped), and destination.

Approach control will issue wind and runway conditions, except when the pilot states, *"Have information Bravo, etc."* This is a phrase used by the pilot to indicate that he has received the automatic terminal information service (ATIS). Approach control will specify the time or geographic position at which the pilot is to contact the tower on local control frequency for further landing clearance. When told to contact the tower, radar service is automatically terminated.

Stage II Service (Radar Advisory and Sequencing Service for VFR Aircraft)

The purpose of this service is to adjust the flow of arriving VFR and IFR aircraft into the traffic pattern in a safe, rapid, and orderly manner. After this service is requested and radar contact established, the pilot is directed to fly specific headings and speeds which will provide adequate traffic separation. A "landing sequence" number then may be issued by the controller.

Stage III Service (Radar Sequencing and Separation Service for VFR Aircraft)

The purpose of this service is to provide separation between all participating VFR

and all IFR aircraft operating within the airspace system defined as the terminal radar service area (TRSA). Within the TRSA, traffic information on observed but unidentified "targets" (other aircraft) will be provided to all IFR and participating VFR aircraft.

EMERGENCY PROCEDURES

Because of the freedom and speed of movement an airplane offers, it does not take long to lose one's bearings. It is especially easy to become lost when flying over featureless terrain and disoriented in heavily populated areas where the great number of roads and small towns appear the same.

DISORIENTATION

The seriousness of being lost is influenced by the weather and terrain conditions encountered along the route. It may not be serious to be 30 miles off course over flat terrain in an area with excellent visibility; however, in mountainous terrain with a poor visibility condition, being off course as much as two or three miles may be a serious matter.

The first procedure to follow when lost is to stay on the original heading and watch for recognizable landmarks. Knowledge of the last known position, elapsed time, and approximate wind and groundspeed allow the pilot to at least estimate the distance traveled since the last known checkpoint. Using this distance as a radius, a semicircle should be drawn ahead of the last known position on the chart. Various landmarks sighted on the ground then should be compared with those in the vicinity of the semicircle. The pilot should realize that his position is probably downwind from the original track.

If fuel exhaustion or darkness is impending, a precautionary landing should be made. A smooth field or a road not bounded by powerlines should be chosen. If darkness or fuel exhaustion is not a threat, the pilot should fly toward a town or developed area. A method to identify a town is to look for names painted in large letters on water towers or tops of buildings. If this procedure is not productive, the town may be identified by its shape and other physical characteristics. Patterns made by railroads and highways are an excellent means of identifying a town or city. When identifying any landmarks, it is important that the pilot not jump to conclusions, but instead, positively identify the landmark or checkpoint. Should the name not appear on a water tower or building, then the "rule of thumb" of finding at least three physical characteristics matching those on the sectional chart should be used.

Once the landmark is positively identified, a direct course to the destination airport should be plotted. The pilot should attempt to determine what initially caused the loss of position orientation. A track miscalculation may have resulted due to overestimating or underestimating the drift correction required to remain on course. Also, miscalculations by the pilot using the computer for time-distance checks may have been the cause. Confusion can occur when prominent landmarks on the ground cannot be compared with corresponding landmarks on the chart. It should be remembered, however, that charts are issued every six months and that a great deal of construction of highways, buildings, and other landmarks may have taken place during the ensuing time.

VERY HIGH FREQUENCY DIRECTION FINDING (VHF/DF)

Another means of determining position is by the aid of VHF/DF steer. Many ground facilities, such as flight service stations and towers, are equipped with direction finding equipment which shows the operator the azimuth (direction) of the aircraft by use of VHF transmissions when the "mike" is "keyed" (depressed). When the signal is received by the flight service station, a needle points in the direction of the transmission. It is necessary for the airplane to be transmitting in order for the DF operator to take a bearing and guide the aircraft. The pilot, after contact has been made with the DF station, has only to follow instructions. The skilled operators are able to direct a pilot straight to or away from any desired location by providing him with headings to fly. Although the DF equipment itself does not indicate the range of the aircraft from the station, an operator can estimate the range fairly accurately and quickly by directing the the pilot to fly certain headings.

Figure 8-25 is an excerpt from the AIM showing how DF service is listed in Part 3. In this case, VHF/DF assistance can be obtained by contacting the Greater Southwest Tower. Since they provide radar service at the Greater Southwest Airport (GSW), the DF equipment is used by tower controllers in conjunction with the radar scopes.

The sectional chart excerpt shown in figure 8-26 illustrates how the availability of DF equipment is depicted on sectional charts. The cluster of airport information just to the left of the McAlister Airport, Oklahoma shows that there is a flight service station on the airport and that it has direction finding equipment.

FORT WORTH
§GREATER SOUTHWEST INTL DALLAS–FT. WORTH FLD *IFR* 16NE
                                            FSS: FORT WORTH (DL)
   568 H90/17–35(2) (S–83, T–120, TT–210) BL5,6,8A,9 S3
F12,18,22,30 Ox1,2,3,4 U2 **RVR:** Rnwy 13
  Remarks:1074′ (1679′ MSL) twr 10.5 NM WSW.
  Southwest Tower 120.5 122.7R   **Gnd Con** 121.8
  Radar Services: (BCN)
     Dallas–Ft Worth App Con 124.5 122.7R
     Dallas–Ft Worth Dep Con 125.2 122.7R
     Stage I Ctc App Con 25 NM out on 124.5
     ASR
  ILS[1] 109.5 I–GSW Apch Brg 129°   **LOM:** 219/GS
  Greater Southwest (H) **BVORTAC** 113.1/GSW at fld.
  VHF/DF Ctc twr.
  Remarks: [1]G/S unusable below 768′ MSL. **VOT:** 111.8

Fig. 8-25. Direction Finding Facility Listing In AIM Part 3—Airport Facility Directory

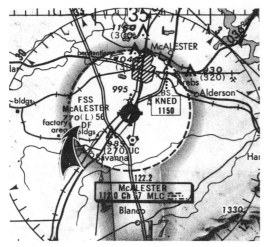

*Fig. 8-26. Direction Finding Listing
On Sectional Chart*

Of the many aids available, radio communications are generally the most useful. Many FAA, military or FCC stations may be contacted on the international emergency frequency (121.5 MHz). When radio contact is made, the pilot should follow the guidelines established by the "Four C s." CONFESS the situation.

All available assistance will be rendered at the pilot's request. He should be ready to COMMUNICATE any or all of the following information: identification, type of aircraft, estimated position, heading, estimated speed, altitude, fuel, nature of the problem, and the assistance desired. It should be fully understood that the facility being contacted may not be able to offer the type of assistance required. Therefore, facility

personnel alert and relay necessary information to those who can provide the needed assistance. If able, the pilot should CLIMB to a higher altitude to provide a greater VHF radio reception distance and, therefore, increase the possibility of DF detection. Finally, the pilot should COMPLY with the instructions of the ground station. Cooperation between pilot and ground personnel is a must for a favorable outcome.

ACCEPTABLE PERFORMANCE FOR CROSS-COUNTRY OPERATIONS

The student pilot will be expected to demonstrate his ability to safely conduct a cross-country flight as the sole manipulator of the controls. He must be able to display complete familiarization with proper preflight action, flight planning, weather analysis, and knowledge of available publications. If requested, the plan for the flight should be explained, calculations verified, and sources of information presented.

During flight, the student must be able to establish and maintain required headings, correctly identify position, and provide estimates of times of arrival within 10 minutes. He must maintain altitude within 200 feet and display the ability to properly divert to an alternate airport, if necessary.

CHAPTER 9
COMMERCIAL FLIGHT MANEUVERS

SECTION A-ACCURACY LANDINGS

Through the accuracy landing, the pilot learns the technique of landing his aircraft when, where, and how he chooses. As the pilot develops the ability to make accuracy landings, he will find that techniques learned in this maneuver will be carried over to his everyday flying.

The accuracy landing, sometimes called a spot landing or precision landing, may also be used in case of an emergency requiring a landing at a place other than an airport. Should the situation *arise*, the pilot, proficient in this maneuver, will be prepared to make a safe, smooth, accurate landing under emergency conditions.

DESCRIPTION OF ACCURACY LANDINGS

The approach to an accuracy landing is normally begun at pattern altitude, although it can be initiated as high as 1,000 feet AGL. The approach to the landing must contain a change in direc-

tion of 180°, and have a *uniform angle of descent* with a touchdown in a normal landing attitude beyond and within 200 feet of a line or predetermined mark. A typical approach to an accuracy landing is illustrated in figure 9-1.

Accuracy landings made by the commercial pilot during the flight test may be made with or without power, slips,

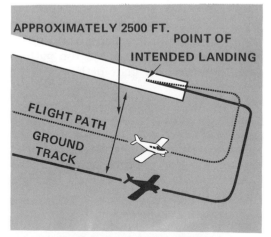

Fig. 9-1. Accuracy Landing

or flaps. Additionally, any combination of these may be used as long as it is within the operating limitations of the aircraft. Many aircraft manufacturers restrict the use of slips with flaps extended.

APPROACH AND LANDING

The consistent performance of accurate landings is the result of proper planning. The pilot must plan the approach so there are as few variables as possible. If the pilot always enters the maneuver from the same altitude, airspeed, and distance from the runway, the glide path will be more uniform and easier to estimate. With these factors constant, the glide path must be altered only to compensate for wind drift.

DOWNWIND LEG

While flying downwind, the pilot should perform the prelanding checklist. Then, while approaching the spot opposite the intended landing point (the 180° point), the pilot should check for other traffic in the pattern that might affect his planning.

The downwind leg should be flown at normal traffic pattern altitude and at a distance of approximately 2,500 feet (one-half mile) from the runway in use. The aircraft's ground track should parallel the runway with no tendency to drift toward or away from the runway. Any deviation from a parallel downwind leg will cause the traffic pattern to have an abnormal shape which greatly influences the length of the pattern and, therefore, the glide path.

At the 180° point, the pilot reduces power to initiate the descent. If a power-off approach is being used, the throttle should be completely closed at this point. For a power approach, the power is reduced as necessary throughout the maneuver.

The pilot should immediately establish normal glide speed by holding altitude

until the normal glide speed is attained. When the approach speed is reached, that speed should be maintained using outside visual references and a descent initiated. The pilot should then continue downwind until in position to turn base leg.

Flaps

Flaps may be used as needed to steepen the landing approach. The recommended procedure would be to add 10° of flaps on downwind and then determine the additional amount of flaps needed while on base leg. Normally, the pilot would add additional amounts of flaps while on base leg and on final approach. Once flaps have been extended, they should not be retracted before the landing touchdown unless the pilot must initiate a go-around.

BASE LEG

The principal indication of when to begin the turn to base leg will come from sighting the aircraft's position relative to the runway. The point of intended landing in a no-wind situation should appear to be approximately 45° behind the wing and look similar to the view shown in figure 9-2 in a standard traffic pattern.

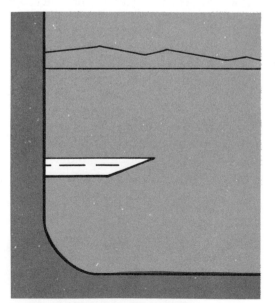

Fig. 9-2. Base Leg Cue

During the turn to base leg, the pilot should use between 20° to 30° of bank. The exact amount of bank used for the turn will depend on wind conditions. To maintain consistency, the same degree of bank should be used for all turns throughout the accuracy landing. Turns with a greater degree of bank than 30° should be avoided within the traffic pattern due to the increased altitude loss.

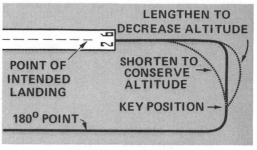

Fig. 9-3. Key Position

Key Position

Once the aircraft turns onto base leg, it will be at the key position. The key position is the point from which the pilot can tell that he is able to glide safely to the field. It is at the key position that the pilot must determine if he needs to conserve or lose altitude. The key position and change in traffic pattern needed to conserve or lose altitude is shown in figure 9-3.

The pilot can do this by a method that has been successfully used by many others. This is the spot method by which the pilot observes his intended point of landing. If the point of intended landing *appears* to be moving down in relation to the aircraft, as shown in figure 9-4, it indicates that the pilot's glide path is high and that he will have to lengthen his base and final legs to lose this additional altitude. If the point of intended landing *appears* to be moving up, the pilot will need to shorten his base and final approach legs by turning toward the runway from this point. As shown in figure 9-5, the pilot may vary the position of the base leg in

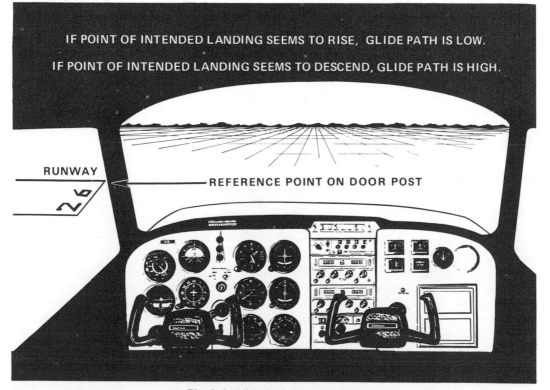

Fig. 9-4. Glide Path Determination

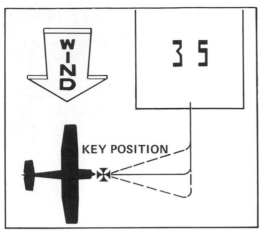

Fig. 9-5. Adjusting Base Leg

order to hold the reference point in a constant relationship to the aircraft.

Glide Estimation

It is at the key position that glide path estimation comes into play. Glide path estimation is made by comparing altitude and distance and making an adjustment for wind velocity. Through experience, a pilot can learn the normal glide path and glide distance for a particular aircraft. Throughout all landing approaches, the experienced pilot will be able to accurately gauge his rate of descent and, therefore, be able to make accurate glide path estimations by comparing the angle of descent with the surface winds.

The pilot will find that a constant airspeed is of utmost importance throughout this maneuver. If the airspeed changes, the rate of descent and thus the glide path will vary considerably. The normal flaps up, power-off glide speed is recommended. This speed provides the pilot with good control feel and a glide path that is between the minimum and maximum glides. The normal glide speed is further recommended because it provides good, solid control throughout the flare.

If the pilot slows the aircraft to the normal glide speed, the shallowest glide path results. If the aircraft is slowed

further, the glide path will steepen considerably. As the aircraft is slowed below the speed which gives the greatest glide distance, the lift/drag ratio causes the aircraft to descend in a steeper glide.

FINAL APPROACH

Before turning to the final approach leg, the pilot should look in all directions for other traffic. If the area is clear, the turn to final can be made.

The pilot must plan the turn to final approach so that the aircraft rolls out on an extension of the runway centerline. The final approach should require *no* sharp angling toward the runway (aircraft too low), nor should it require violent S-turns (aircraft too high). A fairly straight-line approach at the desired airspeed should be maintained through "final" until the beginning of the landing flare.

It can be seen by figure 9-6 that the steeper uniform angle of descent approach provided by use of full flaps is much more accurate than a landing approach without flaps. Because of the steep glide, a 50-foot error in altitude will not cause as large a lateral error in touchdown as a more shallow glide angle.

Slips

Throughout the landing approach, the pilot may use a slip as necessary to lose altitude. The pilot may vary the amount of slip to lose altitude, but the aircraft must be in a normal landing attitude for the actual touchdown.

EVALUATION

The maneuver will be evaluated on the correctness of procedures, airspeed control, coordination, smoothness, and accuracy of the touchdown. The two most important facets of this maneuver are airspeed control and ability to estimate the glide path. The pilot who can maintain accurate control of the air-

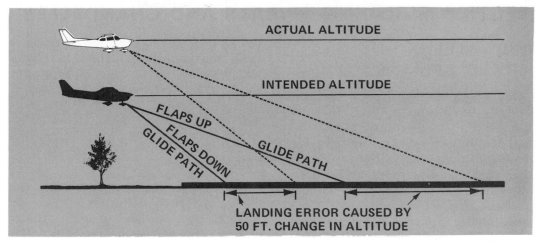

ACTUAL ALTITUDE

INTENDED ALTITUDE

FLAPS UP
GLIDE PATH

FLAPS DOWN
GLIDE PATH

GLIDE PATH

LANDING ERROR CAUSED BY
50 FT. CHANGE IN ALTITUDE

Fig. 9-6. Approach Accuracy

speed will be able to make an accurate glide path estimation.

The aircraft must touchdown in a normal landing attitude slightly above a stall speed beyond and within 200 feet of the anticipated point. Violent maneuvering while making this approach or excessive drift while landing will be disqualifying. Three accuracy landings may be required during the FAA Commercial Flight Test.

COMMON ERRORS

FIXED OBJECT AS KEY POSITION

Many pilots have a tendency to use a fixed object on the ground as the key position. When landing at a different field or making an emergency landing, this fixed object will not be present and the pilot will find that he may be unable to accurately determine the key position. Therefore, the key position should be considered a point in the air a certain distance and angle from the point of intended landing.

LACK OF AIRSPEED MAINTENANCE

Some pilots concentrate so intently on other facets of the landing that they consequently do not maintain a constant airspeed. Since a constant airspeed is a necessity in this maneuver, the pilot must maintain airspeed as closely as

possible. The pilot should maintain airspeed control with reference to outside points and then occasionally cross-check against the airspeed indicator. In this manner, the pilot can monitor his airspeed as he keeps track of the other important considerations.

UNCOORDINATED TURNS

The kinesthesia (sense of coordination) that the pilot developed when performing steep 720° turns will be of great value during this maneuver. The pilot will be able to sense the aircraft movements and tell whether or not the controls are coordinated. An occasional check of the ball in the turn coordinator will confirm the pilot's feel of coordination.

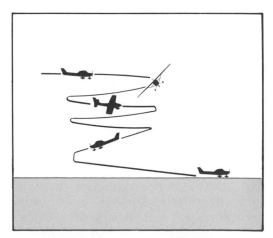

Fig. 9-7. Profile of Gliding Spiral

SECTION B-GLIDING SPIRALS AND CHANDELLES

1080° STEEP GLIDING SPIRALS

Gliding spirals are valuable for teaching coordination, planning, precise speed control, and avoidance of disorientation. Although gliding spirals are learned primarily as a coordination exercise, they also have their practical aspects, such as descending to a landing approach from an altitude higher than normal traffic pattern. If the pilot experiences a power failure at high altitudes, this maneuver provides an excellent procedure for an approach to an emergency landing. Figure 9-7 illustrates an aircraft performing a spiral glide to a point of intended landing.

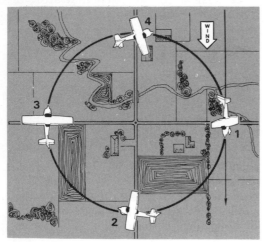

Fig. 9-8. Wind Drift Correction

By practicing this maneuver, the pilot improves his orientation under difficult circumstances. In addition, since precise speed control is one of the most important elements of this maneuver, the pilot mastering it develops an ability to control speed which carries over into all other aspects of flying.

DESCRIPTION

The 1080° gliding spiral is nothing more than a continuous steep bank through three complete gliding turns. During the spiral, a uniform radius is maintained about a reference point on the ground. An intersection of roads or fields provides the pilot with an excellent reference point. To compensate for wind drift, the angle of bank should vary as necessary; however, the pilot uses at least 50° of bank at the steepest point in each turn.

PROCEDURE

Before attempting this maneuver, the pilot must have sufficient altitude to make at least one gliding spiral (1080° of turn) and *recover higher than 1,500 feet above the ground.* Since the aircraft will be descending throughout this maneuver, it is important the pilot clear the area below his aircraft as well as around it. The safest and most positive way to accomplish this is to perform clearing turns and observe the affected airspace as the entry position is approached.

ENTRY

Although this maneuver can be entered from any position relative to the wind, it is recommended that the pilot enter it downwind, approximately one-quarter mile from the point around which the maneuver is flown. By entering the maneuver downwind, the steepest angle of bank occurs at the beginning of the maneuver, thus allowing the pilot to readily adjust the radius of turn as needed to circle the reference point. Entering the maneuver one-quarter of a mile from the selected point generally positions the aircraft so a bank of at least 50° may be used at the steepest point of each turn.

As the pilot approaches the entry point, he should choose a prominent landmark downwind of his position to aid in orientation both during the maneuver and the rollout. Through the use of outside reference points for both pitch attitude and direction, a smooth accurate maneuver and recovery can be performed. Upon arrival at the entry point, carburetor heat is applied, throttle closed, and a gliding spiral begun by rolling into the desired angle of bank. (See Fig. 9-8, position 1.)

AIRSPEED

The speed recommended for this maneuver by the Federal Aviation Administration is 150 percent of the stall speed of the aircraft. This speed provides the pilot with a safe margin above stall speed and a solid, responsive feel of the controls throughout the maneuver.

Since power is reduced to idle, airspeed is controlled solely by the pitch attitude of the aircraft. As the pilot rolls into the bank, he must lower the nose of the aircraft to maintain the correct airspeed. The pilot should establish a pitch attitude that provides him with the speed desired and then maintain that pitch attitude by reference to outside visual references. Reference to the airspeed indicator is used only as a cross-check to confirm the pitch attitude. The pitch attitude in a gliding spiral to the left is illustrated in figure 9-9.

WIND DRIFT

Since the pilot must maintain a constant radius about the selected point, proper wind drift correction is essential. As in the "turns-about-a-point" and "S-turns" the pilot learned while training for his private pilot license, the proper radius is maintained by varying the wind drift correction angle and the angle of bank. On the downwind side of the turn, the bank is the steepest since groundspeed is the greatest. On the upwind side of the turn, the bank is the shallowest. In figure 9-8, the aircraft at position 1 is downwind and consequently has the steepest bank. From position 1 to position 3, the bank angle of the aircraft gradually decreases from the steepest to the shallowest point as it turns upwind. From position 3 to position 1, the bank grad-

Fig. 9-9. Pitch Attitude In Gliding Spiral

LOAD FACTOR	INCREASE IN NORMAL STALL SPEED
1.0	0%
1.5	22%
2.0	41%
2.5	58%
3.0	73%
3.5	87%
4.0	100%

Fig. 9-10. Load Factors

ually increases until it is once again at the steepest point at position 1.

Since this maneuver is entered from a relatively high altitude, the point is more difficult to see and therefore to hold at the beginning of the maneuver. As the aircraft descends, the pilot finds that the constant radius around the reference is easier to maintain.

LOAD FACTOR

An aircraft in a *constant altitude* turn with 60° of bank experiences a load factor of two Gs. Since the gliding spiral is not a constant altitude maneuver, there is less than two Gs of load factor exerted upon the aircraft. However, it is possible for the load factor to increase significantly as the result of the pilot's control usage. If the pilot lets the airspeed increase above normal and then slows the aircraft by raising the nose, the load factor exerted upon the aircraft will increase. Raising the nose of the aircraft tends to tighten the turn rather than raising the nose and adjusting altitude and airspeed; therefore, the load factor can greatly increase with little altitude adjustment.

The increase in stall speed which will accompany the increased load factor can cause an accelerated stall. As shown in figure 9-10, an aircraft with a two G load factor has a stall speed 41 percent greater than normal stall speed. With a three G load factor, the stall speed is increased by 73 percent. For example, an airplane with a normal *indicated* stall speed of 53 miles per hour will stall with a two G load factor at 75 miles per hour; with a three G load factor, it will stall at 92 miles per hour.

As the maneuver progresses, the pilot must not become lax in his search for other aircraft. A constant vigilance for other aircraft in the area must be exercised at all times.

LANDING APPROACH

If the pilot experiences a power failure, he can use a gliding spiral to position the aircraft for an actual landing approach. The pilot finds that it is easier and safer to lose altitude over the point of intended landing than to execute S-turns or other maneuvers that are performed away from the landing area. In the event of a power failure, the pilot should fly his aircraft directly to a point over his intended landing field and then begin the gliding spiral.

The spiral approach also provides the pilot with an opportunity to inspect his intended landing field before he actually commits himself to final approach. Once the aircraft is on final approach, it is a little late for a pilot to change his mind because of obstacles and try to reach another field.

The pilot should plan the spiral approach so he will recover approximately 1,500 feet above ground level *heading*

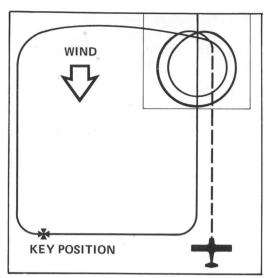

Fig. 9-11. Gliding Spiral To Landing Approach

upwind. At this point, a 360° overhead approach is entered. A spiral approach to a landing pattern is illustrated in figure 9-11. The final approach direction is chosen by the pilot during the gliding spiral (if he does not know the surface wind direction) since he has the opportunity to gauge wind direction and make estimations of his final approach path.

RECOVERY

Normal recovery from a gliding spiral should be made at 1,500 feet above ground level. This maneuver should never be flown lower than 1,000 feet AGL.

The recovery should be made without a change in airspeed when the straight glide is resumed. As the pilot rolls out of the bank to his original entry heading, he should adjust the pitch attitude of the aircraft to maintain the airspeed.

The pilot will discover that adequate practice is the solution to rolling out with the proper pitch control and coordination that is necessary.

The amount of rudder used for recovery will be more than used for entry. Since this maneuver is performed without power the pilot will find the amount of rudder required on rollout is nearly the same for both right and left turns.

EVALUATION

Evaluation criteria is based on drift correction, airspeed control, coordination, orientation, and vigilance for other aircraft. The performance of this maneuver with proper drift correction, airspeed control, and coordination indicates that the pilot learned the elements necessary for the performance of subsequent precision maneuvers.

During the flight test, the pilot is expected to control his airspeed within nine m.p.h. of that desired and his bank between 45° and 55° at the steepest point. Also, his recovery heading must be within 10° of the entry heading.

COMMON ERRORS

AIRSPEED CONTROL

Probably the most common difficulty encountered during steep gliding spirals is proper maintenance of airspeed. Generally, the pilot concentrates too heavily on wind drift control and the bank of the aircraft and finds himself constantly raising or lowering the nose of the aircraft. As the nose of the aircraft is raised of lowered, the airspeed will change appreciably. Through the use of a sighting point on the windshield (as used with previous maneuvers), the pilot can maintain a more uniform pitch attitude and therefore control the airspeed.

ORIENTATION

Since the pilot is flying the aircraft in a steep bank with three consecutive turns, there is a tendency to become disoriented. If possible, the pilot should pick a prominent landmark in the direction of the original heading and then count the number of times he passes it. The pilot greatly improves his orientation by maintaining visual reference to objects outside the aircraft.

Another way to maintain orientation is to use shallower banks and maintain a larger radius of turn during the early periods of training in this maneuver. Once the pilot becomes adjusted to spirals of a more gentle nature, he can progress to the normal steep gliding spirals.

POOR COORDINATION

Due to a lack of propeller slipstream, the student may not use the correct amount of rudder. A sense of coordination must be developed without the propeller blast. As the pilot gains familiarity with the maneuver, coordination will greatly improve.

Another factor contributing to poor coordination is the changing angle of bank required when the maneuver is performed with a wind. Since the bank angle varies (with wind) from approximately 50° to that of a shallow bank, the pilot must vary rudder pressure to remain coordinated. The sense of feel developed in previous maneuvers will be of great value to the pilot during gliding spirals.

CHANDELLES

While learning to perform the chandelle accurately, the pilot will develop a high degree of coordination, planning, control feel, and speed sensing. Through this maneuver, the pilot learns the relationship between control pressures and aircraft attitudes. This precision maneuver incorporates all techniques learned in the previous maneuvers.

DESCRIPTION

The chandelle is a maximum performance climbing turn of 180° duration. Throughout the maneuver, the aircraft speed is smoothly adjusted from entry speed to a few miles per hour above the stall speed by controlling the pitch attitude. As airspeed is reduced, full power is smoothly added and a 180° climbing turn is executed. Figure 9-12 illustrates the performance of a chandelle.

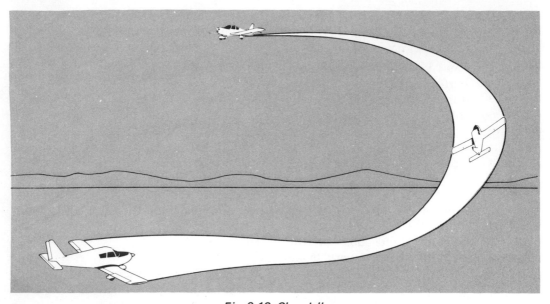

Fig. 9-12. Chandelle

PROCEDURE

Although prevailing winds have little or no effect on this maneuver, the pilot will find it best to begin this maneuver crosswind. By turning into the wind, the smallest amount of wind drift results; therefore, the pilot will find it easier to remain in the practice area. As with all training maneuvers, the pilot should make the necessary clearing turns and check the training area for other traffic.

ENTRY

The chandelle should be entered from level flight at cruising speed. If the cruising speed of the aircraft is higher than maneuvering speed, the aircraft should be slowed to maneuvering speed, or the recommended entry speed, whichever is less. The sequence of events and the relationship between pitch angle and bank angle is illustrated in figure 9-13.

When the chandelle is properly performed, the increase in load factor on the aircraft is extremely small. The load factor encountered in this maneuver should not exceed approximately 1-1/2 Gs; however, by entering the maneuver at or below maneuvering speed, the load factor should not exceed that for which the aircraft is stressed. Maneuvering speed is the maximum speed at which the pilot may use abrupt control travel. Good operating practice in turbulent air is to slow the aircraft to maneuvering speed, or lower, to reduce stress on the aircraft.

As with most precision maneuvers, visual references outside the aircraft are the primary aid for the pilot in maintaining orientation and precision control. Therefore, the aircraft should be aligned with a section line or a prominent landmark to begin the maneuver. While on the desired heading, the pilot should begin a coordinated roll to the desired bank. Although from 20° to 30° of bank may be used in this maneuver, 30° is recommended since it provides the pilot with a smoother, easier maneuver.

FIRST 90° OF TURN

After the bank is established, the pilot applies back pressure to the elevator control to begin the climbing turn. As the climbing turn is initiated, the pilot should gradually add full power in an attempt to maintain cruise r.p.m. as long as possible. If the pilot applies

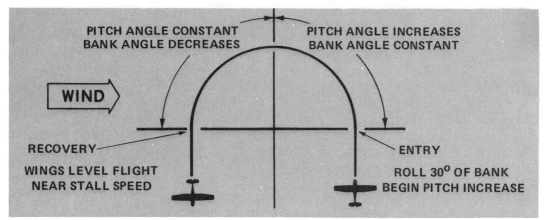

PITCH ANGLE CONSTANT
BANK ANGLE DECREASES

PITCH ANGLE INCREASES
BANK ANGLE CONSTANT

WIND

RECOVERY
WINGS LEVEL FLIGHT
NEAR STALL SPEED

ENTRY
ROLL 30° OF BANK
BEGIN PITCH INCREASE

Fig. 9-13. Plan View of Chandelle

power in this fashion, he will never exceed the maximum r.p.m. for the engine. Throughout the first 90° of turn, the bank of 30° must be maintained. Also, the pitch attitude should be increased at a *constant rate* throughout the first 90° of turn.

SECOND 90° OF TURN

At the completion of the first 90° of turn, the aircraft should still be in a 30° bank and have the highest pitch attitude of the maneuver.

Throughout the second 90° of turn, the pilot should maintain the pitch attitude and slowly roll out of the bank at a *constant rate* until the 180° point is reached. To maintain a constant pitch attitude throughout the second 90° of turn the pilot needs to slowly increase back pressure. As the aircraft loses speed, the elevators become less effective and, therefore, more elevator pressure (control deflection) is needed to maintain the established pitch attitude.

The rollout should be timed so the wings are level at the 180° point. This is accomplished by *reducing the bank at a constant rate* after the 90° point. As the speed of the aircraft decreases and the power is increased, the left-turning tendency caused by P-factor and the propeller slipstream is more prevalent. During the second 90° of turn the rudder pressure needed will be

considerably different in right and left turns.

Since the left-turning tendency at the high angle of attack encountered in the second portion of the chandelle is so evident, the amount of right rudder needed in a chandelle to the right is considerably greater. In a chandelle to the left, the pilot finds that as he passes through the 90° of turn point, he will need to start using *right* rudder to maintain coordination. As the aircraft approaches the 180° point in a left chandelle, the pilot finds that he needs a considerable amount of right rudder to maintain coordination.

RECOVERY

Recovery is completed by lowering the nose to level flight attitude and then increasing airspeed while maintaining altitude.

EVALUATION

The chandelle will be evaluated on the basis of planning, airspeed control, coordination, smoothness, and orientation. Since coordination elements learned in this maneuver are required when performing pylon eights and lazy eights, (the next precision maneuvers), a high-level of proficiency should be attained.

The chandelle must be completed within 10° of desired heading and at an airspeed within five knots of stalling speed. The amount of altitude gained in this maneuver is not the measure of quality. The pilot should exact from his aircraft the best climb performance consistent with the proper use of bank and existing conditions of flight.

COMMON ERRORS

LACK OF COORDINATION

The most common problem experienced by the pilot is lack of coordination. Coordination is generally good until the airspeed starts to decrease and left-turning tendency becomes effective. Generally, insufficient right rudder is the major coordination problem. Control pressures needed for coordination vary greatly throughout the maneuver as the angle of attack changes and, therefore, the left-turning tendency changes.

IMPROPER BANK

The pilot should roll into the desired amount of bank at the beginning of the maneuver and maintain it until 90° of turn, at which point a constant rate rollout is started. The pilot should not be so engrossed with other facets of the maneuver that he forgets about the bank angle.

Some pilots have a tendency to initially roll into a bank that is too steep, resulting in a loss of performance. The pilot will find that with too steep a bank, the aircraft will turn more rapidly and arrive at the recovery point before the airspeed is slowed to the desired speed, or the pilot may find that he is rolling all the bank out in the last 10° to 15° of turn. Also, during the turn with an excessive angle of bank, lift that would otherwise be turned into altitude gain will be used to offset the increased bank.

As the airplane is placed in a climbing attitude during the first 90° of turn, it will appear to the pilot that the bank is increasing; therefore, the pilot will have a tendency to shallow the bank. The attitude indicator may be checked occasionally to confirm the angle of bank. It should be emphasized that the rollout from the 90° point to the 180° point should be at a constant rate and the aircraft should arrive at the recovery point in wings-level flight.

PITCH ANGLE

The pitch angle should increase at a constant rate from the entry to the 90° of turn point. At this time, the pitch angle stabilizes and remains constant throughout the second 90° of turn. With proper planning, the pilot will initiate a pitch angle that cause the aircraft to arrive at the 180° point of turn at a speed slightly above stall speed. With a little practice and the use of outside visual references, the pilot will be able to select a pitch angle that provides the desired results.

If the pitch angle of the aircraft is too great, the possibility of stalling the aircraft before reaching the recovery point exists. With too small a pitch angle, the pilot finds that he arrives at the recovery point with an airspeed greater than five knots above stalling speed. The pilot should remember that since full power is used throughout the last portion of this maneuver, the airspeed is controlled solely by the pitch angle of the aircraft. Because of this, the maintenance of the proper pitch angle is of utmost importance.

SECTION C-PYLON EIGHTS AND LAZY EIGHTS

PYLON EIGHT

The elements learned in pylon eights will aid the pilot in development of subconscious control of the aircraft. Since the pilot's attention throughout this maneuver is directed to visual references outside the aircraft, he must be able to sense the control movements and coordination necessary to align the aircraft with a ground reference point. The pilot who performs excellent pylon eights will gain confidence in his ability to handle the airplane while his attention is diverted outside the cabin.

PREREQUISITES

Since the pylon eight is considered the most advanced and difficult of the low altitude flight training maneuvers, it should not be attempted until the pilot acquires the abilities which are prerequisites for this maneuver. To perform this maneuver properly, the pilot must have the ability to make coordinated turns without gain or loss in altitude. He must have an excellent feel of the aircraft and the ability to completely relax at the controls. In addition, the pilot must be able to concentrate on several items at one time. The student who develops the ability to accurately execute 720° turns, accuracy landings, gliding spirals, and chandelles will be properly prepared for this maneuver.

DESCRIPTION

The pylon eight consists of circling alternately right and left about two reference points on the ground at such a precise altitude and airspeed that a line extending laterally from the pilot's eyes appears to pivot around the reference points. The ground track of the airplane in this maneuver will scribe a figure eight on the ground.

The student must learn to perform both shallow and steep pylon eights. The shallow pylon eights shall not exceed a medium bank at the steepest point. The pylon eight is illustrated in figure 9-14.

SELECTING THE REFERENCE POINT

The reference point, on or near the wingtip, is chosen so the pilot's line of sight through the reference point parallels the lateral axis of the aircraft, as shown in figure 9-15. The pilot's line of sight must also be perpendicular to the longitudinal and vertical axis. The distance of the reference point below the wingtip is dependent on the height of

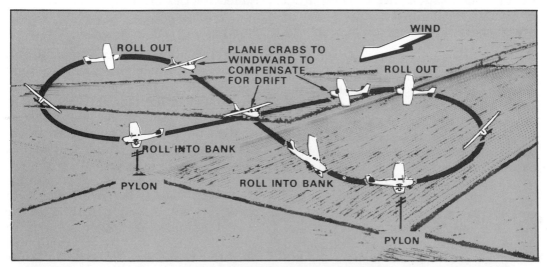

Fig. 9-14. Performing The Pylon Eight

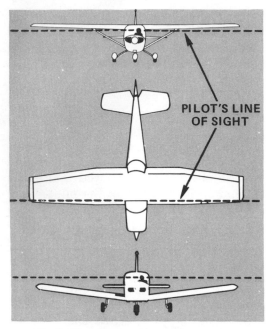

PILOT'S LINE
OF SIGHT

Fig. 9-15. Selecting The Reference Point

the pilot and how near his eye level is to the underside of the wing while sitting in the aircraft.

The best way to determine the reference point is for the pilot to sit in the aircraft on the ground and alternately look out the left and right windows of the aircraft. In this way, the pilot can see that his line of sight is parallel with the ground and the reference point is the same distance from each wingtip. The pilot should fix the reference point in mind in relation to a distance from the wingtip and in relation to a rivet or other discernible mark on the aircraft wing.

SELECTING PYLONS

One of the prime considerations in the selection of pylons is how well they can be seen and identified. Road intersections and fenceposts or telephone poles at intersections make excellent pylons because they can be identified with the least amount of attention. Intersections of fields or objects on the ground may also be used. When rolling out of a turn from one pylon, the next pylon must be readily identifiable so the pilot can plan and prepare for the next turn in the opposite direction.

The pylons should be spaced to provide three to five seconds of straight-and-level flight between the turns. This period of straight-and-level flight serves two purposes. First, it gives the pilot an opportunity to watch for other traffic within the area, and second, it provides the pilot with a chance to properly position the aircraft for entry into the next turn.

Since the turns must be executed at a precise altitude, the pylons should be located on level ground. A variance in altitude between the pylons will necessitate a climb or descent in the straight flight between pylons to position the aircraft at the proper altitude.

Pylons should be selected in open areas and not on the less side of hills or obstructions. Since this maneuver is performed at a relatively low altitude, obstruction induced turbulence, as well as updrafts and downdrafts caused by uneven terrain, could cause the pilot great difficulty in the execution of this maneuver. On hot, sunny days the pilot should also avoid pylons that will cause him to fly over dissimilar terrain, such as plowed fields and fields with lush vegetation. These areas will have local updrafts and downdrafts that will disturb the precise attitude control required for this maneuver.

PIVOTAL ALTITUDE

When the aircraft is in turning flight, there is an altitude that seems to cause the aircraft to pivot about a point on the ground rather than turn about it. It is most noticeable if the pilot attempts to align the reference point with an object on the ground. The height of this *pivotal altitude* is governed by the aircraft groundspeed. At any altitude above this height, the pylon seems to move forward. At any height below the pivotal altitude, the pylon appears to move to the rear. To be performed properly, this maneuver is flown at the exact pivotal altitude of the aircraft.

The groundspeed of the aircraft is the major factor in determining pivotal altitude. As groundspeed is increased, the pivotal altitude also increases. Therefore, the pilot must attempt to use the same airspeed for entering this maneuver as he uses during the maneuver. Throughout, the groundspeed may be changed *very slightly* to alter the pivotal altitude and, therefore, *hold* the pylon.

The *approximate* pivotal altitude is computed by the following formula:

$$\frac{(TAS \text{ in } MPH)^2}{15} = \text{pivotal altitude}$$

For example, if an aircraft is flown at a true airspeed of 100 miles per hour, the pivotal altitude will be approximately 667 feet above the ground.

Use of this formula presents the pilot with an *approximate pivotal altitude* which is very close to the *actual pivotal altitude*. Since existing wind changes the groundspeed of the aircraft, the pivotal altitude also changes slightly, necessitating an adjustment in altitude throughout this maneuver.

Determining the correct pivotal altitude is graphically illustrated by climbing the aircraft to an altitude slightly higher than the altitude determined by the preceding formula and establishing a medium bank turn around a pylon. At this point, the pilot begins a slight descent about the pylon at the airspeed he will use during the maneuver. As the aircraft descends, the pylon moves ahead of the reference point; however, as the aircraft approaches the pivotal altitude, the pylon appears to move to the rear and intercept the reference point. At this point, the pilot must add power to attain level flight while *maintaining the desired airspeed*.

PROCEDURE

The area should be visually cleared to insure no conflicting traffic is present,

since this maneuver is executed at the same relatively low altitude as ground reference maneuvers.

ENTRY

The throttle should be adjusted to a setting which is sufficiently high to handle the maximum angle of bank to be used throughout the maneuver. This should be done as the pilot approaches the entry point, and establishes the airspeed he will use. Also, during this portion of flight, the aircraft should be at or very near the pivotal altitude.

When the airplane is directly opposite the pylon, the pilot immediately lowers one wing to establish the desired bank and align the reference point with the pylon. The sight picture the pilot will see is illustrated in figure 9-16.

Fig. 9-16. Holding The Pylon

HOLDING THE PIVOTAL ALTITUDE

Due to surface winds, the groundspeed varies at different points around the pylon and causes the pivotal altitude to vary slightly. While flying downwind, the groundspeed is the greatest and, therefore, the pivotal altitude is the highest. Conversely, when flying upwind at the slowest groundspeed, the pivotal altitude is the lowest.

Throughout the turn around the pylon, the pilot must gradually but constantly adjust his altitude to hold the reference point on the pylon.

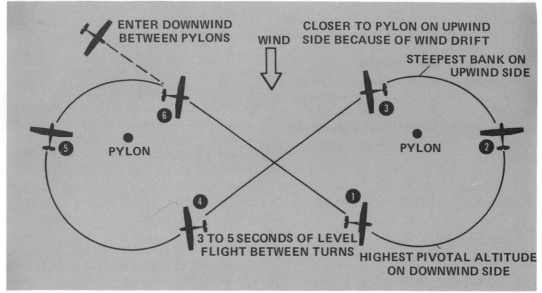

Fig. 9-17. Plan View Of Pylon Eight

The aircraft at position 1 & 4 in figure 9-17 has the greatest groundspeed and, therefore, the highest pivotal altitude. As the aircraft progresses to position 2 or 5, the groundspeed is the slowest and, therefore, the pivotal altitude the lowest. As the aircraft proceeds around the pylon, the groundspeed begins to increase with a resulting increase in pivotal altitude until the aircraft reaches position 3 or 6. At this point, the aircraft is rolled into straight-and-level flight.

If the pylon appears to move to the rear of the reference point, as shown in figure 9-18, the pilot should increase altitude slightly. This action has a double effect upon the pivotal altitude. First, the aircraft will climb to the higher pivotal altitude necessary to hold the pylon, and second, the climb will cause the airspeed to decrease which also lowers the pivotal altitude to meet the climbing airplane.

If the pylon seems to move forward, as shown in figure 9-19, it indicates the aircraft is too high and the pilot must descend to the pivotal altitude. This also has a double correction effect since the process of descending to the pivotal altitude increases airspeed which, in turn, will raise the pivotal altitude.

Fig. 9-18. Pylon Moves Rearward

Fig. 9-19. Pylon Moves Forward

The following formula will aid the pilot in remembering which control movement is needed under each condition.
1. pylon forward, control wheel forward
2. pylon rearward, control wheel rearward

WIND DRIFT AND BANK ANGLE

Drift correction is used only during the straight portions of flight between the pylons. While flying around the pylons, the pilot is concerned with holding the reference point on the pylon and does not attempt drift correction. Because of this, the aircraft will be the closest to the pylon while on the upwind side of the maneuver. In this position, the wind is drifting the aircraft closer to the pylon, and the pilot must have a steeper bank to hold the reference point on the pylon. The drift and bank angle of the aircraft in this wind condition is illustrated in figure 9-20.

It must be noted that bank angle of the aircraft has no effect on pivotal altitude. The varying degrees of bank necessitated on the downwind and upwind sides will have little effect on pivotal altitude, since this height is determined by groundspeed and not the bank angle. The only way that bank angle can affect pivotal altitude is for bank to increase to such a degree that it causes a decrease in groundspeed.

EVALUATION

Performance is based on planning, altitude control, coordination, smoothness,

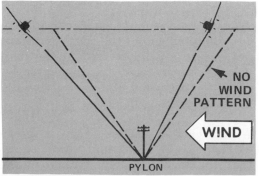

PYLON

Fig. 9-20. Wind Drift And Bank Angle

and the ability to *hold* the pylons. With proper planning and altitude control, coordination and smoothness will follow and provide for proper execution of the maneuver. Throughout this maneuver, the pylon must be held within one foot of the line-of-sight reference.

COMMON ERRORS

LACK OF COORDINATION

The most common error is the lack of coordination caused by the use of rudders. Many pilots try to hold the reference point on the pylon by use of rudder instead of holding the pylon by changing altitude and airspeed. As the pylon moves to the rear, a common error the pilot makes is to hold the pylon by applying bottom rudder pressure. This tends to yaw the wing back. As with other precision maneuvers, the rudder is used only to trim the aircraft to a balanced flight condition. Use of rudder pressure to hold the reference point on a pylon results in a slip which is a disqualifying maneuver.

Although it is helpful to check the ball in the turn coordinator occasionally, the pilot should develop a *feel* (kinesthesia) for the aircraft which will enable him to maintain proper coordination. By developing the *feel* of the aircraft while learning maneuvers discussed previously, the pilot will overcome the need to continually monitor the inclinometer and therefore will not find it necessary to divert his attention from the pylons.

FAILURE TO HOLD THE PYLON

Failure to hold the pylon is generally caused by the pilot's inability to select and maintain the pivotal altitude. The pilot who is having difficulty holding the pylon will find that spiraling down to and intercepting the pivotal altitude will be a beneficial exercise. This interception practice will assist the pilot in determining the exact pivotal altitude.

It must be emphasized that the pilot needs only a small amount of altitude correction to hold the reference point on the pylon. The change in altitude between the downwind portion of the pylon eight to the upwind portion may be so slight that the pilot will not detect the difference in the altimeter indication.

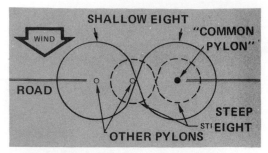

Fig. 9-21. Picking The Pylon

The most accurate way to realign the reference point with the pylon is to note the trends of movement and then anticipate the control pressures that are needed. For example, if the pylon begins to move rearward, the pilot raises the nose slightly, increasing altitude and decreasing airspeed. At the point the pylon seems to stop moving to the rear, the pilot starts relaxing the control pressure. The pylon will move forward and as realignment is noted, the pilot should establish the pitch attitude that provides level flight. If, instead of anticipating the control movement necessary, the pilot waits until the pylon is reintercepted before relaxing control pressure, the pylon will move past the reference point in the opposite direction.

DRIFT CORRECTION BETWEEN PYLONS

The length of the straight flight segment between pylons varies in accordance with the amount of bank desired around the pylon. For example, on steep pylon eights, the straight portion of flight must position the aircraft closer to the next pylon than on shallow pylon eights. Therefore, when the pilot rolls out of the turn between pylons, he must adjust the straight segment of flight to compensate for wind drift and place the aircraft in a proper postion for the next turn. The pylons may be selected closer to each other for the steep pylon eight, as illustrated in figure 9-21.

LAZY EIGHTS

The lazy eight is a training maneuver that combines dives, climbs, turns, and various combinations of each. Through this maneuver, the pilot continues to develop his coordination, speed sense, and subconscious feel of the aircraft. During lazy eights, control pressures are constantly changing, necessitating careful advance planning of control usage to perform the maneuver well. Because of this constant control pressure change, the lazy eight cannot be done mechanically or automatically. The flight path of an aircraft performing a lazy eight is illustrated in figure 9-22.

DESCRIPTION

The lazy eight is essentially two 180° turns in opposite directions, one following the other, with each turn having a climb and a dive. It is called a lazy eight because the longitudinal axis of the aircraft appears to scribe a figure eight about the horizon. Figure 9-23 shows how an extension of the longitudinal axis appears to draw this eight on the horizon.

To execute a smooth, precision maneuver, all turns within the lazy eight must not exceed a bank of 30°. If the pilot uses a steeper bank, the maneuver becomes hurried and looses the smoothness that is desired.

A plan view of the lazy eight is shown in figure 9-24. This illustration shows that the highest pitch attitude in the climb comes after 45° of turn, and that

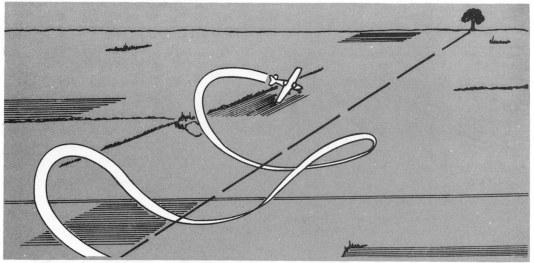

Fig. 9-22. Flight Path Of Lazy Eight

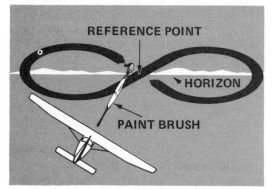

Fig. 9-23. Painting The Lazy Eight

the lowest pitch attitude in a dive comes after 135° of turn. During the first 90°, the bank angle is increasing at a constant rate until it reaches 30°. Throughout the second 90° of turn, the bank angle is decreasing constantly until wings-level flight is reached after 180° of turn. The lazy eight is a beautiful maneuver, and for the pilot who masters it, a very gratifying experience.

EFFECTS OF WIND

The turns of a lazy eight should be made into the wind; if this is not done, the loops of the lazy eight will not appear to be of equal size. If the maneuver is executed crosswind, the wind will make the loops of the lazy eight cross the horizon at different points and the longitudinal axis of the aircraft will draw an unsymmetrical eight about the horizon. Making turns into the wind will also tend to keep the aircraft within the training area.

PROCEDURE

As with all of the training maneuvers, the pilot must first clear the area for other traffic. During the lazy eight, the aircraft will be changing directions and altitudes constantly, and the pilot must first clear the area and then maintain vigilance throughout the maneuver for other traffic.

The pilot should choose a reference object on the horizon as a center point for the eight that is scribed about the horizon. This reference object should be the point that the longitudinal axis of the aircraft passes through at the center of the figure eight. A section line, building, or possibly a tall tree on the horizon are all good reference objects. In any case, the reference object should be directly upwind of the aircraft and aligned with the wingtip at the start of the maneuver.

ENTRY

This maneuver should be entered from straight-and-level flight at cruising speed, maneuvering speed, or recommended entry speed, whichever is lowest. The power setting used for this ma-

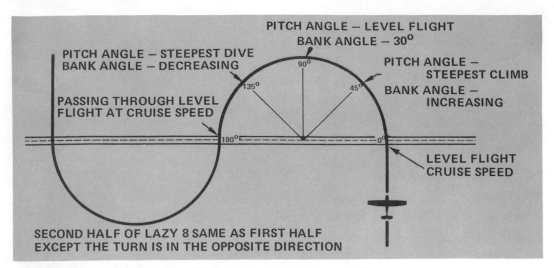

Fig. 9-24. Plan View Of Lazy Eight

neuver will normally be cruise power; however, the pilot should be able to select a power setting that will not cause a persistent gain or loss in altitude.

FIRST 90° OF TURN

Figure 9-24 illustrates the sequence of events that occur throughout the lazy eight. The pilot begins the lazy eight by initiating a gentle climbing turn. Through the first 45° of turn, the bank angle is increasing slowly, and the pitch angle is increasing to maximum.

At the 45° point of turn, the aircraft is at the highest pitch attitude and the bank is still increasing. As the aircraft passes through the 45° of turn point, the pilot begins to gently lower the nose so the aircraft will reach a level flight pitch attitude after 90° of turn. Throughout the first 90° of turn, the bank angle is increasing at a constant rate and is timed so the desired bank of 30° is reached at the 90° of turn point.

SECOND 90° OF TURN

As shown in figure 9-24, the aircraft is in level pitch attitude and 30° bank after 90° of turn. It should be noted that as the aircraft passes through level flight the extended longitudinal axis intersects the reference object.

After passing through the first 90° of turn, the pilot initiates a constant rate rollout timed so the aircraft will complete the first 180° of turn in a wings level attitude. After passing the first 90° of turn, the pilot continues decreasing his pitch attitude until he reaches the lowest point after 135° of turn and the bank is approximately 15° to 20°.

As the nose of the aircraft passes through 135° of turn, the pilot begins gently raising the nose to increase the pitch attitude. Through the last 45° of turn, the pitch attitude rises to level flight and the bank angle continues to decrease.

RECOVERY

Recovery should be timed so the bank angle reaches 0° and the pitch attitude reaches level flight after 180° of turn. In addition, the aircraft should recover at the entry speed and altitude.

The same procedure is used throughout the next 180° of turn in the opposite direction. As the aircraft reaches level flight after 180° of turn, the pilot immediately begins the climbing turn in the opposite direction. The airplane is not flown in straight-and-level flight, but only passes through straight and level at the 180° point of turn as it is rolled into the next turn.

EVALUATION

Performance is based on planning, orientation, coordination, smoothness, altitude, and airspeed control demonstrated by the pilot. Since the attitude, altitude, and speed of the airplane are constantly changing, the pilot must display a high degree of piloting skill in the proper performance of this maneuver. Repeated slipping or a persistent gain or loss in altitude at the completion of the maneuver will be disqualifying factors.

COMMON ERRORS

IMPROPER PLANNING

The pilot should plan this maneuver so the peaks of the loops both above and below the horizon come at the proper location within the maneuver. The peak of the loop above the horizon should come at approximately 90° of turn, and the lowest altitude should come as the aircraft is passing through straight-and-level flight. Additionally, the pilot must plan for the changing bank angle which is used throughout the maneuver. Therefore, proper attention to bank angle, pitch angle, and aircraft heading becomes absolutely necessary.

The pilot can aid his orientation and planning by dividing each 180° of turn into four segments of 45° each. Pre-planning the events in each 45° segment insures complete understanding and makes proper anticipation a simple matter.

UNSYMMETRICAL LOOPS

When properly performed, the peak of a loop above the horizon is approximately the same size and shape as the peak below the horizon. Through proper airspeed control, the pilot can make symmetrical loops in this maneuver.

Pitch attitude in the climbing turn during the first portion of the maneuver must provide a speed slightly above stall as the aircraft passes through level flight. In similar fashion, the pitch angle in the diving turn must allow the aircraft to accelerate to entry speed after 180° of turn. If these two criteria are adhered to, loops will be approximately symmetrical and equal in size.

TOO STEEP A PITCH ATTITUDE

If the pilot uses too steep a pitch attitude in the climbing turn, the aircraft may stall before reaching the 90° point. The nose of the aircraft should pass through the reference point after 90° of turn at the minimum maneuver speed. This speed is normally a small amount above stall speed.

Too low a pitch attitude in the second portion of the turn results in an excessive dive which causes the aircraft to exceed entry speed at the 180° point. The excessive pitch attitude with a resultant gain in airspeed causes the pilot to lose altitude in the maneuver and to enter the second half of the lazy eight at the wrong airspeed.

TOO STEEP A BANK

Since the maneuver seems easier to perform with steeper than medium banks, some pilots have the tendency to steepen the bank angle beyond normal. The steeper bank will cause the pilot to hurry through the maneuver with a resulting lack of precise control. The lazy eight should be performed as a slow, lazy maneuver with only 30° of bank at the steepest point.

ALPHABETICAL INDEX

a

ACCELERATED STALLS, 5-6
  acceptable performance, 5-8
  load factors, 5-6
ADVANCED MANEUVERS, 4-1
AILERON LOCKS, 1-24
AIR NAVIGATION, METHODS OF, 8-21
  checkpoints, 8-22
  dead reckoning navigation, 8-21
  flying the VFR trip, 8-22
  pilotage, 8-21
  radio navigation, 8-21
AIR WORK, 8-9
AIRCRAFT AIRWORTHINESS CERTIFICATE, 1-2
AIRCRAFT CONTROL, 7-3
AIRCRAFT LOGBOOKS, 1-2
AIRCRAFT REGISTRATION CERTIFICATE, 1-2
AIRPLANE RADIO STATION LICENSE, 1-2
AIRPLANE TIEDOWN, 1-22
AIRPORT LIGHTING, 8-5
  rotating beacons, 8-5
  runways, 8-5
  taxiways, 8-5
AIRSPEED INDICATOR, 2-9
AIRWORTHINESS CERTIFICATE, 1-2
ALTITUDE CORRECTION RULES, 7-6
ALTITUDE, PIVOTAL, 9-14
ANGLE OF ATTACK, 4-8
ANTICOLLISION LIGHTS, 8-3
APPROACH-TO-LANDING STALLS, 4-7
  causes, 4-8
  certification regulations, 4-7
  Part 23, 4-7
APPROACHES, 3-12
ARRIVAL, 8-26
ATTITUDE FLYING, 2-1, 2-3, 2-8
ATTITUDE INDICATOR, 2-1, 2-8
ATTITUDE INSTRUMENT FLYING, 7-1
  aircraft control, 7-3
  altitude correction rules, 7-6
  bank control instruments, 7-7
  climbing and descending turns, 7-13
  cross-check, 7-2
  descents, 7-10
  instrument climbs, 7-9
  instrument interpretation, 7-3
  level-off from climb, 7-9
  level-off from descent, 7-11
  pitch control, 7-5
  power-on spiral recovery, 7-14
  psychological factors, 7-1
  straight-and-level flight, 7-3
  turns, 7-11
  unusual attitudes, 7-13

b

BANK CONTROL INSTRUMENTS, 7-7
BASE LEG, 1-27, 3-16
BEST ANGLE-OF-CLIMB SPEED, 2-11, 5-9
BEST ANGLE-OF-GLIDE SPEED, 2-13
BEST RATE-OF-CLIMB SPEED, 2-11, 5-10
BOUNCED LANDINGS, 3-25

c

CABIN FAMILIARIZATION, 8-4
CENTRIFUGAL FORCE, 2-18
CHANDELLE, 9-9
CHECKPOINTS, 8-22
CLIMBING AND DESCENDING TURNS, 7-13
CLIMBING TURNS, 2-26
CLIMBS, 2-10
  best angle-of-climb speed, 2-11
  best rate-of-climb speed, 2-11
  cruising climb speed, 2-11
  normal climb speed, 2-11
CONSTANT-RADIUS TURNS, 6-7
CONTROL LOCK, 1-2
CONTROL SURFACE LOCK, 1-3
COORDINATION EXERCISES, 2-28
  advanced coordination exercise, 2-30
  dutch roll, 2-28
COWL FLAPS, 1-5
CROSS-CHECK, 7-2
CROSS-COUNTRY OPERATIONS, 8-15
  arrival, 8-26
  departure, 8-22
  emergency procedures, 8-29
  enroute, 8-23
  enroute weather information, 8-24
  flying over a cloud layer, 8-25